Hydrostatic, Aerostatic, and Hybrid Bearing Design

Hydrostatic, Aerostatic, and Hybrid Bearing Design

W. Brian Rowe DSc, FIMechE

Emeritus Professor of Mechanical Engineering,
Liverpool John Moores University

and

Consulting Engineer, Court Cottage, Moult Hill,
Salcombe, Devon, TQ8 8LF, UK

ELSEVIER

AMSTERDAM • BOSTON • HEIDELBERG • LONDON • NEW YORK • OXFORD
PARIS • SAN DIEGO • SAN FRANCISCO • SINGAPORE • SYDNEY • TOKYO
Butterworth-Heinemann is an imprint of Elsevier

Butterworth-Heinemann is an imprint of Elsevier
The Boulevard, Langford Lane, Kidlington, Oxford, OX5 1GB
225 Wyman Street, Waltham, MA 02451, USA

First published 2012

British Library Cataloguing in Publication Data
A catalogue record for this book is available from the British Library

Library of Congress Number: 2012932618

ISBN: 978-0-12-396994-1

For information on all Butterworth-Heinemann publications
visit our website at www.elsevierdirect.com

Printed and bound in the United States

12 13 14 15   10 9 8 7 6 5 4 3 2 1

Contents

Preface

The special qualities of pressurized hydrostatic, aerostatic, and hybrid bearings afford a simple and convenient solution to many bearing problems experienced with particular machines. Sometimes, the best method of achieving a specified performance is to use a hydrostatic bearing, an aerostatic bearing, or a hybrid bearing. However, the designer is not always experienced in hydrostatic and aerostatic lubrication, and has difficulty obtaining authoritative guidance presented in a simple manner.

This book has been written with this problem in mind and is based on the author's personal experience over many years in bearing design and testing, in running courses on this subject for industry, and in writing articles for the technical press.

Theory is kept to an elementary level in the design sections and the book offers a useful introduction for engineers who have left academic study behind. A basic theory chapter refreshes relevant principles of fluids. The book is also useful to students of engineering design and lubrication. There is growing awareness in universities and colleges of the importance of tribology for reliability and effectiveness of all moving devices and mechanisms. This book covers important sectors of thin-film lubrication and machine design.

Acknowledgements are due to many valued colleagues, particularly Dr J. P. O'Donoghue, Dr K. J. Stout, Dr D. A. Koshal, Dr F. S. Chong, Professor K. Cheng, Professor W. Weston, and Dr D. Ives, main colleagues with whom analysis, computing, and experimental investigation were developed. A considerable volume of experimental work undertaken over the years has contributed to understanding where theory fits with reality and, more importantly, where it diverges.

Great care has been paid to ensure accuracy in the design guidance given in this book. But, however carefully the text has been checked, some readers may find ambiguity. The technique of presenting and explaining principles wherever possible, followed by procedures and examples, allows the reader to cross-check intentions and overcome such problems.

W. Brian Rowe

Consulting Engineer and Emeritus Professor of Mechanical Engineering
Liverpool John Moores University
2012

Usual Meaning of Symbols

Suffix	Meaning
a	Axial
ae	Aerostatic
c	Value for capillary control
d	Diameter, diaphragm control, or discharge
e	Effective
f	Friction (as in friction power)
h or hb	Hybrid
hd	Hydrodynamic
hs	Hydrostatic
i	Inner or sometimes inlet
j	Journal
m	Maximum condition
o	Design value or concentric value as in h_o
o	Orifice value as in A_o
p	Pumping as in H_p
r	Condition at bearing recess or sometimes radial value
s	Supply value as in P_s or slot value as in a_s
t	Total as in H_t or axial thrust as in A_t, or transition as in N_t
v	Valve
1, 2	First, second recess or bearing, etc.
overbar	Dimensionless value, for example $\overline{A}$ or $\overline{B}$

Symbol	Dimensionless Symbol	Meaning
a		Land width in main flow direction
a_s		Width of restrictor slot
A		Projected total bearing area
A_e	$\overline{A} = A_e/A$	Effective bearing area
A_f	$\overline{A}_f = A_f/D^2$ or $\overline{A}_f = A_f/L^2$	Friction area for journals / Friction area for flat pads
A_r		Recess area
A_t		Total bearing area in sliding contact
b		Inter-recess land width
b		Source groove or source hole separation
	$\overline{B} = \dfrac{q\eta}{p_r h^3}$	Shape factor for flow rate
B		Bearing width
c		Specific heat capacity or width of axial slot in slotted journal bearing
C		Damping
C_d		Orifice discharge coefficient
C_{kp}		A factor for flow through an orifice
C_p		Specific heat at constant pressure
C_v		Specific heat at constant volume
c_{vr}		Periphery around a virtual recess
d_c		Diameter of capillary
d_o		Diameter of orifice
D		Journal or other diameter
e	$\varepsilon = e/h_o$	Journal eccentricity
E		Opposed pads area ratio ($E = A_{e1}/A_{e2}$)
f_n		Natural frequency in hertz
F		Force
h	$\overline{X} = h/h_o$	Bearing lubricant film thickness
h_o		Clearance or design film thickness
H	$\overline{H}$	Power dissipation
H_f		Friction power
H_p		Pumping power
H_t		Total power
i		Set: $i = 1, 2, 3, \ldots$
j		Set: $j = 1, 2, 3, \ldots$
J		Mechanical equivalent of heat
k		Polytropic exponent in gas law
K		Power ratio ($K = H_f/H_p$)
K		Bulk modulus of liquid
K_p		Absolute pressure ratio ($K_p = p_r/p_s$)
K_c		Capillary factor $\left(K_c = \dfrac{128 l_c}{\pi d_c^4}\right)$
K_{go}		Concentric gauge pressure ratio $\left(\dfrac{p_{ro} - p_a}{p_s - p_a}\right)$

K_s		Slot factor $\left(K_s = \dfrac{12l_s}{a_s z_s^3}\right)$
l_c		Capillary length
l_s		Slot length
L		Bearing length
L		Lower clearance ratio limit, $2h_o/D$
m		Mass or mass-flow rate for gas bearings
n		Number of recesses
n_s		Number of slots per row
N		Rotational speed in rev/s
N_o		Rotational speed at which $K = 1$
p	$\bar{p} = p/p_s$	Absolute pressure
p_a		Absolute ambient pressure
p_r		Absolute recess pressure
p_{ro}		Absolute concentric recess pressure
p_s		Absolute supply pressure
P	$\bar{P} = P/P_s$	Gauge pressure $P = p - p_a$
P_r	$\bar{P}_r = P_r/P_s$	Gauge recess pressure
P_{ro}		Concentric gauge recess pressure
P_{av}	$P_{av} = W/LD$	Average pressure on journal bearing
P_s		Constant gauge supply pressure at supply source
q	$\bar{Q} = \dfrac{q\eta}{P_s \bar{B} h_o^3}$	Volume flow
q_a		Aerostatic flow in terms of volume of free air or gas
q_r		Flow through a restrictor
QF		Approximate load and flow factor for dispersion losses
r		Radius
R		Perfect gas constant
S_h		Hybrid speed parameter
$S_h = \dfrac{\eta L U}{P_s h_o^2} = \sqrt{\dfrac{K\beta \bar{B}}{\bar{A}_f}}$		S_h for flat-pad bearings
$S_h = \dfrac{\eta N}{P_s}\left(\dfrac{D}{C_d}\right)^2 = \dfrac{1}{4\pi}\sqrt{\dfrac{K\beta \bar{B}}{\bar{A}_f}}$		S_h for journal bearings
$S_{ho} = \sqrt{\dfrac{\beta \bar{B}}{\bar{A}_f}}$		S_h for flat pads when $K = 1$
$S_{ho} = \dfrac{1}{4\pi}\sqrt{\dfrac{\beta \bar{B}}{\bar{A}_f}}$		S_h for journals when $K = 1$
SF		Extent of source surround of a virtual recess
t		Time
T		Thermodynamic temperature
T	$\bar{T}$	Axial thrust force

(Continued)

—Cont'd

Symbol	Dimensionless Symbol	Meaning
ΔT		Maximum temperature rise
u, v		Journal coordinates (v along line of eccentricity)
U		Surface speed
U		Upper clearance ratio limit, C_d/D
w		Small film force additional to static value
W	$\overline{W} = \dfrac{W}{P_s A_e}$	Static bearing film force
W		Radial force for journal bearings
x, y, z		Coordinates or small displacements
	$\overline{X} = h/h_o$	Bearing lubricant film thickness
Z		Impedance (pressure difference/flow rate)
z_s		Slot restrictor film thickness
α (alpha)		Semi-cone angle
$\beta = P_{ro}/P_s$ (beta)		Gauge pressure ratio when $h = h_o$
γ (gamma)		Circumferential flow factor for journal bearings
γ (gamma)		Ratio of specific heats C_p/C_v
δ (delta)		Orifice restriction ratio
Δ (delta)		Prefix denoting an incremental value
$\varepsilon = e/h_o$ (epsilon)		Eccentricity ratio
η (eta)		Damping ratio (C/C_c)
ζ (zeta)		
θ (theta)		Angle of inter-recess land
$\Theta = x/w$ (theta)		Transfer function of bearing film
λ (lambda)		Bearing film stiffness
ν (nu)		Kinematic viscosity
ρ (rho)		Density of the bearing lubricant
τ (tau)		Shear stress or time constant
ϕ (phi)		Attitude angle between W and e in a journal bearing
ω (omega)		Frequency of excitation in rad/s

CHAPTER 1

Application

1.1 Introduction

Hydrostatic and aerostatic bearings have a great attraction to the engineer because machine parts supported on such bearings move with incomparable smoothness. This apparent perfection of motion derives from the complete separation of the solid sliding surfaces with a fluid film. At no point do the solid surfaces make any physical contact. This means to say the thin fluid film separating the surfaces is always larger than the height of any surface irregularities and as a result there is a complete absence of sticking friction. A mass supported on a hydrostatic or aerostatic bearing will silently glide down the smallest incline, an effect which is most striking with very large machines.

The hydrostatic bearing appears to have been invented by Girard (1852), who employed high-pressure water-fed bearings for a system of railway propulsion. Since that date there have been hundreds of patents and publications dealing with different designs and incorporating novel features. While some of these designs are potentially useful, many others introduce complexity rather than simplicity and are destined to remain in the archives. This text concentrates on configurations attractive for effectiveness and simplicity.

Aerostatic bearings employ pressurized air or gas whereas hydrostatic bearings employ a liquid lubricant. Aerostatic bearings offer some advantages but compressibility of gas imposes limitations in design layout compared with hydrostatic bearings. This book unifies the analysis of aerostatic and hydrostatic bearings, thus simplifying the design process and making it possible for the designer to achieve a wider range of designs.

Gauge Pressure and Absolute Pressure Conventions

The following convention is adopted throughout the book: *The upper case letter **P** always refers to gauge pressure—that is, the excess of pressure above ambient pressure as measured on a typical pressure gauge:*

$$P = p - p_a$$

*The lower case letter **p** refers to absolute pressures—that is, the excess of pressure above absolute zero.*

Hydrostatic, Aerostatic and Hybrid Bearing Design.
DOI: 10.1016/B978-0-12-396994-1.00001-2

1.2 What are Hydrostatic, Hybrid, and Aerostatic Bearings?

The Hydrostatic or Aerostatic Principle

In a hydrostatic or aerostatic bearing the surfaces are separated by a film of fluid forced between the surfaces under pressure. The pressure is generated by an external pump. Hence a more general term "externally pressurized bearing" is often used. However, the separate terms "hydrostatic" and "aerostatic" are used here to maintain a clear distinction and allow differences to be made clear. An advantage of both types of bearing is that a complete lubricant film is maintained whenever the bearing is pressurized, even at zero speed.

Hydrostatic bearings should not be confused with hydrodynamic bearings, where although pressure is employed, the pressure does not support the applied load.

Hydrostatic and aerostatic circular pad bearings, with orifice flow control, are shown in Figure 1.1. Lubricant at a constant supply pressure P_s is pumped towards the bearing. The pressurized lubricant first passes through the orifice where dissipation of pressure energy causes reduced pressure on entry into the recess of the bearing pad. The recess is relatively deep compared with the bearing film thickness so that it offers little resistance to flow. Pressure in the recess is therefore constant throughout the recess volume. The flow passing through the recess leaves through the thin gap between the bearing land and the opposing surface. The pressure in the bearing film reduces as it passes across the bearing land and reaches

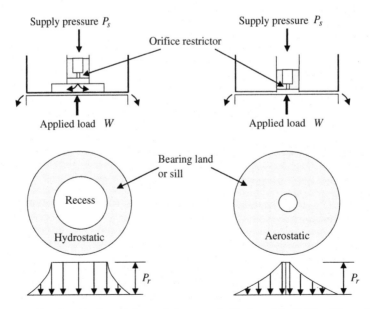

Figure 1.1: Circular Hydrostatic and Aerostatic Pads with Orifice Flow Control.

atmospheric or ambient pressure at the exit. Other types of flow-control device can also be employed, such as the laminar flow group including capillary and slot restrictors.

The film pressures oppose the applied load and maintain the separation of the surfaces. Recess pressure must be lower than supply pressure to allow for load variations. This is because the recess pressure must be able to vary with applied load. The principle can be demonstrated by two extreme cases. In the first case, a high applied load forces the bearing surfaces together and prevents flow out of the bearing. The flow rate through the orifice decreases to zero. Recess pressure therefore rises until it equals supply pressure. The second extreme is when load is reduced to zero. In this case the bearing gap will become very large so that the only resistance to flow is that offered by the orifice restrictor. This causes the flow to increase until the pressure drop across the orifice is sufficient to reduce the recess pressure to ambient pressure. The permissible range of applied loads must be such that the film thickness remains between the two extremes.

The load supported is calculated from the pressures in the bearing. Figure 1.2 shows two examples of hydrostatic pressures based on a simplified longitudinal flow assumption. The flow restrictors shown are assumed to be slots that reduce the inlet pressures P_i to a value equal to $\frac{1}{2}P_s$. The bearing film force is therefore $W = \frac{1}{4}P_s LB$. In the second example, a recess allows the inlet pressure P_i to spread uniformly throughout the recess. For a recess of length b and width B, the contribution from the recess to the load is $P_i bB$. The total load support includes the contribution due to the triangular distribution on the lands of length l. Each of these lands contributes a load support of $\frac{1}{2}P_i lB$. The total bearing film force or load support is therefore $W = P_i lB + P_i bB = \frac{1}{2}P_s lB + \frac{1}{2}P_s bB$. The principle for aerostatic bearings is similar.

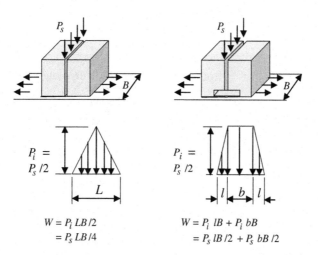

Figure 1.2: Examples of Hydrostatic Load and Pressure in One-Dimensional Longitudinal Flow.

The distinction between typical hydrostatic and hydrodynamic bearings is illustrated in Figure 1.3. In the hydrostatic example, lubricant enters four recesses through separate entry ports. When the shaft is concentric, pressures are almost constant around the shaft. Restrictors in the supply lines to each recess allow pressures to vary when the shaft is not concentric.

The concentric value of recess pressure, at zero load, is usually half the supply pressure, that is $P_{ro} = P_s/2$. If the bearing gap on one side is reduced, the bearing gap on the other side of the bearing is increased. Flow through the smaller gap is reduced and recess pressure rises. If the shaft is completely displaced to one side of the bearing, the recess pressure is almost equal to the supply pressure. On the opposite side, flow is increased and recess pressure drops. The reduction in pressures on one side is accompanied by an increase on the other. Thus, both sides of the bearing contribute force to withstand the externally applied force on the shaft. The main load parameters are supply pressure and bearing area.

   W is proportional to $P_s LD$ for hydrostatic and aerostatic load support.

Two main configurations for hydrostatic, aerostatic, and hybrid journal bearings are illustrated in Figures 1.4 and 1.5. The recessed bearing is suitable for hydrostatic operation. Large recesses reduce friction area and power consumption at speed. Plain slot-entry and hole-entry bearings are suitable for all three modes of operation. Plain bearings minimize gas volume in an aerostatic bearing and hence reduce the effects of compressibility. Plain bearings maximize hydrodynamic support in hybrid bearings.

Slot-entry aerostatic bearings can be designed with a reasonable degree of certainty. In contrast, orifice-fed aerostatic bearings introduce a degree of uncertainty due to: (1) dispersion losses and (2) the possibility of pneumatic hammer instability. It is always good practice to test a prototype bearing to ensure that the design requirements have been met.

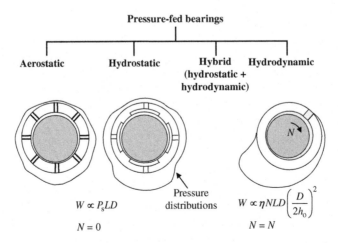

Figure 1.3: Aerostatic, Hydrostatic, Hydrodynamic, and Hybrid Journal Bearings.

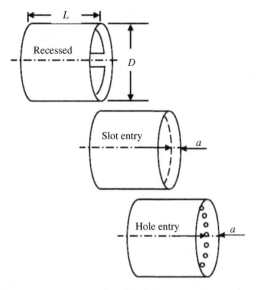

Figure 1.4: Recessed and Plain Journal Bearings.

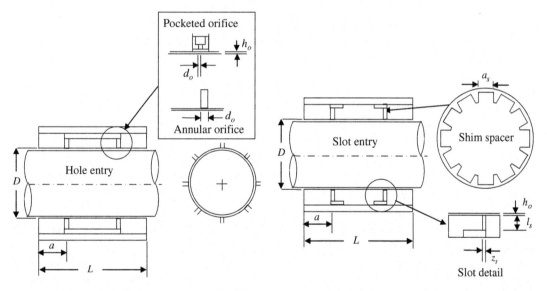

Figure 1.5: Typical Hole-Entry and Slot-Entry Hydrostatic, Aerostatic, and Hybrid Bearings.

The Hydrodynamic Principle

The hydrodynamic bearing illustrated in Figure 1.3 is said to be "self-acting" because the hydrodynamic pressures are generated by movement of the bearing surfaces. The moving surface drags lubricant by means of viscous forces into the converging gap. The converging gap region occurs on one-half of the bearing between the maximum gap on one side and the minimum gap on the other. The result is pressure in the converging region. The resulting bearing film force is equal and opposite to the applied force on the shaft. Hydrodynamic load support is primarily dependent on speed, viscosity, and bearing area.

W *is proportional to* $\eta NLD(D/2h_o)^2$ *for hydrodynamic load support.*

At zero speed, $N = 0$, there is no bearing film force and no bearing film load support. This is the condition on starting when external forces are supported by physical contact between the two bearing surfaces. Physical contact leads to wear. For this reason, starting and stopping are the main causes of wear in hydrodynamic bearings. Hydrostatic bearings should not experience wear since pressure is switched on before start-up and ensures complete physical separation of the bearing surfaces.

Hybrid Hydrostatic and Hybrid Aerostatic Bearings

At speed, applied forces on a hybrid hydrostatic bearing are supported by the combined effect of hydrodynamic and hydrostatic bearing film pressures. Hybrid bearings are designed to take maximum advantage of hydrostatic and hydrodynamic effects. Attractive features include zero start-up wear, good bearing film stiffness at zero speed, and high load capacity at full speed. Hybrid aerostatic bearings employ the same principle combining aerostatic and aerodynamic contributions.

Two plain hybrid bearings are illustrated in Figure 1.5. Recesses in the surfaces are avoided in order to maximize hydrodynamic load support. Hybrid hydrostatic or hybrid aerostatic bearings may be conveniently designed with slot-entry or hole-entry restrictors.

The Aerostatic Bearing Principle

In principle, an aerostatic bearing operating with an externally pressurized air or gas supply is similar to a liquid hydrostatic bearing. Gas is fed through a restrictor and escapes through the gap between the bearing lands. However, compressibility of a gas means that a prototype orifice-fed aerostatic bearing should be tested to avoid the risk of bearing instability. Other types of restrictors such as annular orifices, slot, and capillary reduce the risk of instability, although at the expense of lower concentric stiffness. Average stiffness over the whole load range is much the same for different types of restrictors.

One of the main differences between aerostatic bearings and hydrostatic bearings is that recesses for aerostatic bearings are very small in area and shallow in depth to limit the response time between application of load and change in recess pressure. A long response time can lead to an instability known as "pneumatic hammer". Aerostatic bearings with large recesses do not always demonstrate instability, but the probability is greatly increased.

Since large recesses are usually avoided, aerostatic bearings are similar in layout to hybrid hydrostatic bearings. In fact, Dee and Porritt (1971) demonstrated that bearings of the slot-entry type can be made interchangeable for gas, liquid, or steam applications.

1.3 When are Hydrostatic, Hybrid, and Aerostatic Bearings Employed?

Most hydrostatic bearings require an additional pump to be incorporated into the machine in order to supply lubricant under pressure to the bearing. While a pump is normally required for a hydrodynamic bearing, the pumping pressure is often very low and the pump is likely to be inexpensive. A higher pressure supply system is an increased cost for a hydrostatic bearing. However, a high-pressure source of lubricant used for another function in the machine may sometimes be capable of supplying the bearings. This may be the case if the machine has hydraulic equipment for actuation, clamping, or spindle drives.

Aerostatic bearings have the advantage that with suitable filtration and moisture traps, air can be used from a centralized high-pressure supply system. This eliminates the cost of a high-pressure supply system required for hydrostatic bearings. An additional safety feature is provision of a high-pressure accumulator that supports the bearing for a limited period in case of electrical supply or air supply failure. An accumulator is important for large high-speed aerostatic bearings.

Space and cost requirements need to be considered at an early stage of design. Such considerations are often of overriding importance for manufacture in large quantities. The converse tends to be more relevant when only one or two special machines are required. In this case, hydrostatic or aerostatic bearings may prove to be less expensive as well as offering superior performance, long life, and reliability.

Another cost consideration is the requirement for control restrictors and effective filtration to prevent blockage in the supply. Cost is relatively less important in a machine where the total cost is very large, where performance and reliability are much more important. Similarly, space and weight requirements are less important if the machine is stationary or very large.

Among the attractive features of hydrostatic and aerostatic bearings is the ability to operate at zero speed and at high speed. In addition, high bearing film stiffness may be achieved. Film stiffness is a measure of the applied load necessary to produce a small change in lubricant film thickness. With special types of flow restrictor, it is even possible to design the stiffness independently of the load. This allows the designer to design the bearing performance to suit the operational requirements of the machine.

Other features are low starting torque, high accuracy of location, good dynamic stability, and cool operation. Another advantage is that selection of materials is usually less critical than with most other bearing types.

In view of these considerations, it is not surprising that hydrostatic and aerostatic bearings have been successfully employed for many years in a number of large low-speed machines that require high load support and low friction in order to achieve high precision in positioning. In the USA, for example, several radio telescopes have been supported on hydrostatic bearings. A notable early example includes the 63 m (210 ft) diameter Goldstone radio antenna.

The machine tool industry has made extensive use of hydrostatic and aerostatic bearings, where they have proved to be reliable and predictable and have often improved the performance of the machine beyond the capability of any other type of bearing. The exceptions have been where insufficient attention has been given to system design, including filtration and the prevention of restrictor blockage.

In machine tools it is important that the bearings are not subject to wear, which makes it impossible to maintain machining tolerances and production rates. Wear also reduces the resistance to chatter in metal-cutting operations. Such features are particularly important for automatic machine tools.

A recent example of successful redesign of a large grind-turn machine used for the production of wind-turbine bearings was reported by KMT Lidköping (Editorial, 2010). The earlier machine was capable of machining 600-mm-diameter bearings while maintaining a form deviation within 3 μm. The new machine employing hydrostatic slides was claimed to be capable of a feed resolution of 0.1-μm steps and producing 4000-mm-diameter bearings to a form deviation of 1 μm.

Other applications include support bearings for experimental apparatus such as bearing test rigs and dynamometers, and also bearings and seals in hydraulic motors where a ready source of pressurized oil is available.

A feature that could be important in some applications is the ability to achieve a strong vibration damping action. This has been employed for noise damping.

The power requirements of the pump are not necessarily large or even proportional to the size of the machine. In fact, the power requirements are a function of the product of speed squared and friction area. A typical example of a journal bearing of 100 mm (4 in) diameter, shaft rotational speed 5 rev/s (300 rev/min) with axial thrust of 9000 N (2000 lbf) requires a power of 40−80 W (0.05−0.10 hp).

Hybrid bearings have not received as wide consideration in the technical and scientific literature as purely hydrostatic or purely hydrodynamic bearings. Hybrid bearings, however, exhibit interesting and useful properties. A plain hybrid bearing when designed appropriately performs as a superior bearing at high speed with the attractive features of a hydrostatic bearing at low and high speeds. These features include good load capacity and stiffness achievable at any speed, improved dynamic stability compared with hydrodynamic bearings, and cool operation. Additional features are the ability to employ higher viscosity oils at high speeds and the tolerance of wider variations in bearing clearance.

The hybrid bearing is superior to both axial groove and circumferential groove hydrodynamic bearings when dynamic load is to be applied in widely varying radial directions. The disadvantage of hybrid bearings is the same as for hydrostatic bearings: it is normally necessary to provide auxiliary hydraulic equipment, effective filtration, and flow control restrictors. However, it is possible that a lower system pressure will suffice in view of the high overload capability.

There are obvious applications where plain hybrid bearings have performance advantages. These are high-speed machines where hydrodynamic journal bearings may suffer from whirl instability at low eccentricity ratios, as in generator sets, turbines, and vertical spindle pumps for large thermal power installations, and also for machine tools subject to intermittent cutting operations, shock loads, and occasional heavy overloads. The bearings can be safely operated at higher eccentricity ratios, reducing the risk of whirl.

A special class of the hybrid bearing is the jacking bearing in which the bearing is jacked hydrostatically under pressure to separate the bearing surfaces and hence avoid wear and high friction under starting and stopping conditions. At speed, the supply pressure may possibly be reduced or even switched off to allow the bearing to operate purely hydrodynamically from a separate low supply pressure.

The application of aerostatic bearings is outlined further in Section 1.6.

1.4 Bearing Selection

Bearing Configurations

Hydrostatic and aerostatic bearings can be designed in a wide variety of configurations, as indicated in Figure 1.6 for hydrostatic bearings. Some particular journal configurations have already been shown. However, the range of different configurations and types of control restrictors requires separate consideration of the operational requirements of each type of bearing. The choice of bearing geometry, bearing size, and control system is considered in more detail below.

While these bearing configurations are also relevant to aerostatic bearings, recesses must be made much smaller and very shallow to avoid pneumatic hammer. Some of the figures indicate the modifications to a hydrostatic bearing required for aerostatic operation.

Bearing Load Support

Hydrostatic Thrust Pads

An approximate guide for a safe mean operating load is

$$W = 0.25 \times \text{supply pressure} \times \text{bearing area}$$

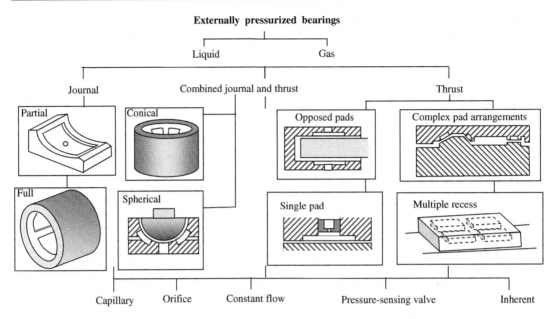

Figure 1.6: Hierarchy of Externally Pressurized Bearings.

The mean load value is usually termed "the design condition". Thus, if a pad is designed to support 500 N (112.5 lbf), it will usually be possible for the pad to support a range of applied forces that either increase the total load by one-third to a maximum of approximately 667 N (150 lbf) or reduce the total load by one-third to approximately 333 N (75 lbf).

Aerostatic Thrust Pads

Aerostatic bearings with central gas admission support much lower loads than hydrostatic pads for the same supply pressure. This is due to the much smaller recesses that are employed. Further guidance on effective pad areas is provided in Chapter 4. Load capacity of aerostatic pads can be improved by surrounding a central land area with a ring of shallow entry holes or slots to create a high-pressure region. Such a region is termed "a virtual recess". However, stability may be undermined and a prototype bearing should be tested.

Journal Bearings

The following is a guide for a well-designed bearing having length/diameter equal to 1.

$$W = 0.25 \times \text{supply pressure} \times \text{diameter squared}$$

Aerostatic journal bearings support similar loads to hydrostatic journal bearings for the same pressure. Since aerostatic bearings generally employ much lower pressures, this limits the loads that can be supported on a given pad area. However, friction power in an aerostatic bearing is very small compared with other fluid-film bearings. It may therefore be possible to employ a much larger diameter journal so that load and stiffness match a hydrostatic bearing.

Plain Hybrid Bearings

Load capacity at zero speed is much the same as for recessed hydrostatic bearings but is substantially improved at higher speeds.

Flow

Flow must increase with sliding speed and with bearing area to ensure adequate cooling in hydrostatic and hybrid bearings. Small pads operate with very small bearing film thickness and low flow since little cooling is required. Large pads require larger film thickness to achieve greater flow for cooling. There is a greater risk of variations in manufactured bearing film thickness with large pads and this reinforces the need for a larger film thickness.

Aerostatic bearings operate with a viscosity 1.82% of the viscosity of water. Cooling is not usually a consideration except at extremely high speeds. However, flow rates for aerostatic bearings tend to be much higher than for hydrostatic bearings and therefore film thicknesses are usually smaller.

Relative Merits of Different Types of Bearing

The relative merits of hydrostatic and other types of bearing are shown in Table 1.1, where a simplified rating system has been adopted. This is a guide where each characteristic is considered in isolation. Where particular characteristics are required in combination, such as high accuracy, high stiffness, and low wear for a machine tool, the relative merits with respect to costs may be reversed since a hydrostatic or aerostatic bearing may be the only bearing where these properties are easily achieved and maintained.

Cost

The initial cost of hydrostatic bearings is likely to be much higher than for an equivalent rubbing bearing. This is because a high-pressure pump system and control restrictors are required for a hydrostatic bearing. However, the life of a hydrostatic bearing is virtually infinite. This results in the machine retaining its initial accuracy and efficiency throughout its life. The film thickness of an aerostatic bearing is very small, so accurate manufacture is vital.

During its long operating life, the bearing can sometimes be run on cheaper lubricants and even on the process liquid itself. Friction losses are often smaller than with other types of bearing, which offsets the power required for the pump. With a correctly designed hybrid bearing, the total power consumption is actually reduced.

Table 1.1: Comparison of bearing types

Operation Factor	Plane Rubbing	Rolling Element	Hydrodynamic (Self-Acting)	Hydrostatic/ Aerostatic	Hybrid (Plain)
Initial cost	****	***	**	*	*
Standard parts	****	****	**	*	*
Space	**	****	**	**	**
Axial	**	****	**	**	**
Radial	****	*	***	***	***
External	Not required	Not required	**	*	*
Load	***	***	****	***	****
Run cost	****	***	**	*	*
Low speed	****	***	*	****	***
High speed	*	**	****	****	****
Starting torque	*	**	*	****	****
Running torque	*	**	***	****	****
Accuracy of position	**	***	**	****	****
Whirl	***	****	**	***	***
External vibration	**	*	**	**** Hs ** Ae	****
Life	*	**	***	****	****
Ease of design	****	**	*	*** Hs ** Ae	***

Aerostatic bearings do not require such a costly supply system as hydrostatic bearings, but cost is increased by the need for larger bearings, corrosion-resistant materials, small clearances, and also by the need to manufacture and fit restrictors.

1.5 Bearing Categories

Hydrostatic and aerostatic bearings mostly fall into the three broad categories illustrated in Figure 1.6: journal bearings, thrust bearings, and combined journal and thrust bearings.

Journal Bearings

A common oil-lubricated journal bearing configuration is the cylindrical bearing containing four, five, six, or more recesses each controlled by its own restrictor. Aerostatic bearings will usually have 8−16 restrictors in a single row configuration or two similar rows in a double-row configuration. Recesses are not an essential feature of hydrostatic journal bearings, although most commercially produced hydrostatic journal bearings do contain recesses. Alternative plain bearing configurations shown in Figure 1.4 may offer advantages for manufacturing simplicity and reliability. Plain bearing configurations are much more suitable than recessed configurations for aerostatic bearings.

Combined Journal and Thrust Arrangements

Examples of combined journal and thrust bearings are illustrated in Figure 1.7. Most journal bearings require axial constraint and a common method is to employ a thrust flange supported between two annular recessed pads. Thrust bearings for journals tend to require relatively high flow rates due to the problems of achieving sufficiently wide bearing lands in the space available. Flow rate may be reduced by designing a calliper arrangement instead of the usual 360° thrust pads. An alternative is the conical configuration, which can withstand radial and axial loads. In this case, it is necessary to decide whether this advantage is worth the extra manufacturing complexity. The Yates bearing achieves radial and axial loads in a simple and economical way. The oil that leaks from a conventional cylindrical journal bearing has to escape through the thrust bearings. It has been found that such a system allows quite substantial axial loads to be supported in addition to radial loading.

The following is a guide with optimal design to safe applied loads. Load T in the direction of the principal rotational axis is applied in combination with load W applied in a radial direction. Guides are based on the spindle diameter D and gauge supply pressure P_s. However, since there are many variables involved, it is advisable to follow a detailed design process for a particular shape and bearing sizes.

These guides apply both for liquid and for gas bearings. For aerostatic bearings, shapes are suitably modified to avoid large and deep recesses. Suitable shapes for aerostatic bearings are discussed in later chapters.

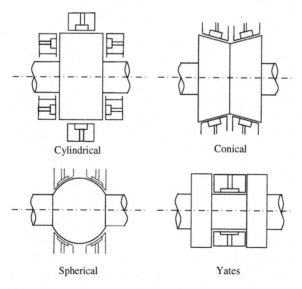

Cylindrical   Conical

Spherical   Yates

Figure 1.7: Combined Journal and Thrust Bearings.

1. Cylindrical journals ($L/D = 1$), $W/(P_sD^2) = 0.25$ for each bearing
   Annular thrust ($D_{max}/D_{min} = 1.25$), $T/(P_sD^2) = 0.17$ for a pair
2. Conical journals, $W/P_sD^2 = 0.2$ for each bearing
   (Semi-cone angle $\alpha = 10°$, $L/D = 1$), $T/(P_sD^2) = 0.12$ for a pair
3. Partial spherical bearings, $W/(P_sD^2) = 0.09$ for each bearing
   (Included angle 50°), $T/(P_sD^2) = 0.06$ for a pair
4. Yates bearing, $W/(P_sD^2) = 0.18$ for each bearing
   ($L/D = 1$ and $D_{max}/D_{min} = 1.25$), $T/(P_sD^2) = 0.08$ for a pair.

In most situations, a spindle requires two journal bearings and a pair of opposed thrust bearings, as shown in Figure 1.8. These arrangements may support an overhanging load at one or both ends of the spindle. The most common arrangement is the combination of two cylindrical journal bearings and a pair of annular recess flat-pad thrust bearings.

The conical arrangement shown in Figure 1.8 has lower resistance to overhanging loads than the conical arrangement shown in Figure 1.7. If the conical bearing surfaces in Figure 1.8 lie tangentially around the arc of a circle, the resistance to overhanging loads will be very poor. The cone angle should therefore be small and the two journals well separated in the axial direction.

The same comment applies to the spherical arrangement in Figure 1.8. The two journals need to be well separated if the bearing is to support overhanging loads.

Most spindle arrangements support ample thrust loads in the axial direction. The Yates bearing shown in Figure 1.8 supports lower thrust loads than the conventional plane thrust pads but provides adequate thrust loads for many applications. The advantage of the Yates bearing is simplicity and a reduced flow rate requirement.

Flat-Pad Bearings

The simplest example of a flat sliding bearing is the circular or rectangular single-recess pad shown in Figure 1.6. However, a single flat pad must always be held down by a positive force. Where the applied load reverses in direction, as in some machine-tool slideway systems, opposed pads are employed.

The thrust pad with a single recess has virtually no resistance to tilt. This may be an advantage in a spherical bearing that is designed to allow free rotation in any direction but, for most machines, pads must be arranged in a suitable pattern to ensure alignment of the bearing surfaces. The rectangular multi-recess bearing pad for a linear slide and the multi-recess annular thrust pad for rotary movement may both be employed where tilt resistance is required. Both types of hydrostatic pad are illustrated in Figure 1.9. For gas bearings the recess size would be greatly reduced both in area and in depth to avoid pneumatic hammer.

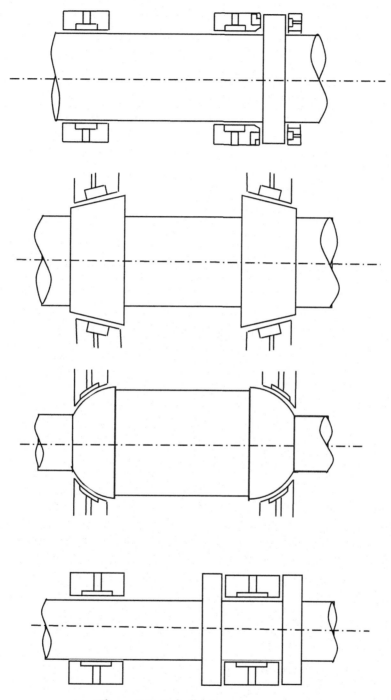

Figure 1.8: Spindle Arrangements.

Rectangular multi-recess   Annular multi-recess

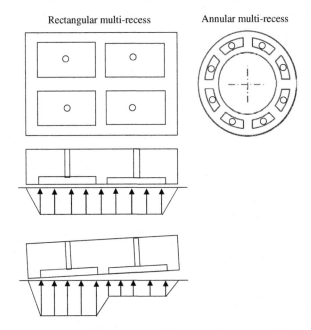

Figure 1.9: Multi-Recess Pads Offer Tilt Resistance.

A multi-recess annular thrust pad of the type shown in Figure 1.9 was used to support the analyzing magnet of the Nuclear Structure Facility at the Science and Engineering Research Council Daresbury Laboratory. This bearing was designed in cooperation between Liverpool Polytechnic and Daresbury (Rowe et al., 1980). The analyzing magnet in position is shown in Figure 1.10.

The magnet bends the beam of ions through 90° towards the experimental equipment. The diameter of the bearing is 1.8 m (72 in); it carries a load of 53 tonnes (52 tons) and its surface is flat to an accuracy of 5 μm (0.0002 in). The operating requirements involved maintaining the vertical centerline of the apparatus within 0.1 mm (0.004 in) radius at a height of 6.3 m (21 ft) above the bearing face, while the magnet may be rotated to direct the beam into any one of three experimental areas. This was achieved with a supply pressure $P_s = 1.1$ MN/m^2 (162 lbf/in^2) and a pressure ratio $\beta = 0.5$. Removable capillary tubes were used for flow control for reasons of simplicity and ease of cleaning.

1.6 Commercial Applications

Hydrostatic, hybrid, and aerostatic bearings are found in a wide range of production machines. Hydrostatic bearings are often employed to support very heavy loads that must be precisely positioned as in machining centers or massive scientific instruments as illustrated in

Figure 1.10: The Analyzing Magnet Supported on a Hydrostatic Bearing at the Base of the Nuclear Structure Facility at the SERC Daresbury Laboratory.

Figure 1.10. Hydrostatic bearings are particularly favored for precision grinding machines where close machining tolerances must be held over long production runs.

For example, Cinetic Landis Ltd manufactures a wide range of CNC controlled grinding machines for processing crankshafts, camshafts, and transmission parts for the automotive industry. The majority of the machines use hydrostatic rotary and linear bearing systems to provide good stiffness, damping, and low error motion. Workpiece sizes range from small motor cycle cams up to large marine diesel crankshafts weighing in excess of 8 tonnes. Figure 1.11 shows the LT1 model incorporating hydrostatic grinding spindles capable of delivering 22 kW of process power at 25,000 rpm.

Zollern GmbH produces a wide range of hydrostatic bearings as components that other manufacturers can incorporate into their products. The range includes spindle bearings, linear bearings, and lead-screw bearings. An interesting and unique feature of some ranges is a patented bearing clearance compensation design that obviates the need for orifices and other control devices. Figure 1.12 shows an example of a conventional Zollern heavy-duty bearing suitable for radial and axial load support.

Westwind Air Bearings Ltd is a longstanding and well-known manufacturer of aerostatic/aerodynamic spindle bearings employed for a range of applications such as dental drills and drills for electronic components. Figure 1.13 shows an air bearing spindle employed for drilling printed circuit boards. The spindle operates at speeds up to 200,000 rpm.

Figure 1.11: (a) Cinetic Landis LT1 grinding machine featuring hydrostatic spindles and linear bearings. (b) A high-speed 100 mm hydrostatic spindle and bearings. *Courtesy of Cinetic Landis Ltd*

Air bearings are employed for two reasons. Very small drills require high rotational speeds to achieve high production rates. Air bearings are ideal for such high speeds. High-speed drills need to be fed very precisely to avoid drill breakage. Very low friction makes air bearings ideal for precise position control.

1.7 Materials and Manufacture

Manufacture and Construction

The machining and construction of thrust pads is usually straightforward. For journal bearings, several methods have been employed for producing recesses. These include milling or

Figure 1.12: Zollern Radial-Axial Hydrostatic Bearing Used for Heavy-Duty Lathes.
System: one pump per pocket. Bearing size: 500 mm diameter × 350 mm. Operating load: 500,000 N. Spindle speed: 200 rpm. Radial and axial runout: <5 μm. *Courtesy of Zollern GmbH*

Figure 1.13: A Westwind Air Bearing PCB Drilling Spindle, with Integral Pneumatic Automatic Tool Change Capability.
Speed range: 20,000–200,000 rpm. Bearing air supply gauge pressure: 5.6 bar. Bearing air flow rate: 40 nl/min. Shaft diameter: 19 mm. Fitted with a water-cooled direct drive 200-volt AC electric motor. Max axial load: 20 kgf. Max radial load at collet: 6.8 kgf. Dynamic runout of collet at 200,000 rpm: <7 μm. *Courtesy of Westwind Air Bearings*

grinding, which is difficult for small and medium-sized bearings of less than 250 mm (10 in) diameter. Other methods include electrical discharge machining (EDM), electroplating, etching, or fabrication. The latter provides a simple method particularly for smaller assemblies. Examples of fabricated journal bearings are illustrated in Figures 1.14 and 1.15. Figure 1.14 shows the parts of a recessed journal bearing and Figure 1.15 the parts of a slot-fed journal bearing for hydrostatic or hybrid operation and either liquid or gas lubrication.

Design and manufacture of flow restrictors is discussed in Chapter 5.

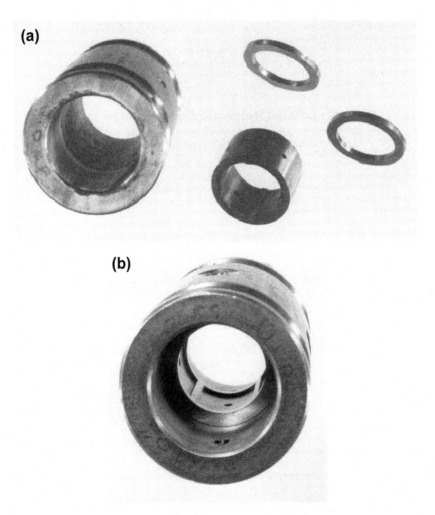

Figure 1.14: (a) Four-recess hydrostatic journal bearing dismantled after 10 years' extensive use. There is a complete absence of burnished or scored zones. The material is steel. (b) Bearing assembled.

(a)

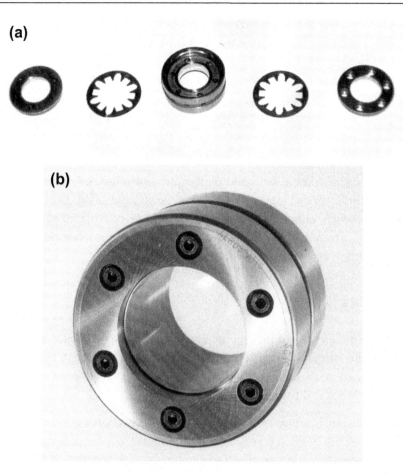

(b)

Figure 1.15: (a) Slot-entry plain journal bearing dismantled after 4 years' extensive use.
There is a complete absence of burnished or scored zones. The material is steel.
(b) Slot-entry bearing assembled.

Materials Selection

The selection of materials for the oil-lubricated bearing surfaces is not as critical for hydro-static bearings as for other bearing types because in normal operation there is no contact between the bearing surfaces. There are circumstances, however, where careful consideration must be given to the materials to be employed. Some of these may be listed as follows:

1. High bearing pressures. Soft bearing materials may extrude or otherwise deform unacceptably.
2. High varying temperatures. Expansion coefficients will be important to maintain clearances and materials must be stable at the high temperatures. This is particularly important with hot-gas or liquid-metal lubrication.

3. Movement when power is disconnected. If the machine is likely to be moved and adjusted with the power disconnected, materials should be selected to avoid any scoring of the bearing surfaces.
4. Extremely high-precision movements. The materials should be selected for dimensional stability and ease of machining.
5. Corrosive lubricants. Nonreactive materials or surfaces must be employed.
6. Air bearings. Expansion of pressurized air inevitably leads to condensation of moisture. Materials therefore need to be corrosion resistant.

The selection of materials should be based on the principles long established in engineering practice. In most applications, materials selection is straightforward. Compatible bearing pairs are normally chosen, although not necessarily, and selected from mild steel, hardened steel, chromed steel, silvered steel, graphite cast iron, brass, phosphor bronze, lead bronze, bearing alloys, ceramics, polymers, and composite materials.

1.8 Aerostatic Bearings

Air bearings have been employed in large numbers for high-speed drills and cutters used in dental and orthopedic surgery. More generally, gas bearings have been employed for larger machines requiring extremely high accuracy such as roundness measuring machines, straightness measuring machines and for smaller grinding machines.

The geometry of aerostatic bearings is similar to plain hybrid/hydrostatic bearings as described in Chapter 10. This is because large recesses must be avoided in aerostatic bearings to prevent low dynamic stiffness and reduce the risk of self-excited pneumatic hammer. Plain slot-entry aerostatic bearings are best for dynamic performance and definitely better than orifice-fed bearings.

Gas-lubricated bearings offer some of the following advantages (Powell, 1970):

*   Extremely low friction, cool running, and low power loss
*   Capable of very high speeds
*   Low or zero wear rate
*   Low noise and vibration levels
*   Capable of very high and very low temperature operation
*   Gases such as air and nitrogen are nonpolluting in the atmosphere.

There are also disadvantages compared with hydrostatic bearings:

*   Supply pressures are usually limited to 4—6 bars and therefore load support is reduced
*   Orifice-fed gas bearings are vulnerable to pneumatic hammer instability.

This book unifies and simplifies the design approach employed for hydrostatic and aerostatic bearings.

1.9 How to Read and Use the Book

It is recommended to read the first six chapters to gain a basic understanding of general principles and their application. The first six chapters contain background principles of importance in design. Afterwards the designer can move to a relevant chapter for a particular design. Design procedures with sample calculations are given in each chapter.

References

Dee, C. W., & Porritt, T. E. (1971). The design and application of the universal slot-fed fluid bearing. *Proceedings of the Southampton University Gas Bearing Symposium, Vol. 1*. Paper 12.

Editorial. (2010). Professional engineering. *Proceedings of the UK Institution of Mechanical Engineers, 23*(11; 7 July), 39—40.

Girard, L. D. (1852). *Nouveau Système de Locomotion sur Chemin de Fer*. Paris: Bachelier.

Powell, J. W. (1970). *The Design of Aerostatic Bearings*. Machinery Publishing Company.

Rowe, W. B., Morris, M. C., & Acton, J. (1980). A note on the design, manufacture and testing of the hydrostatic bearing for the Daresbury Laboratory Nuclear Structure Facility 90° analyzing magnet. *Tribology International, 13*(2), 183—184.

CHAPTER 2

Basic Flow Theory

2.1 Introduction

The two most important parameters are bearing flow and fluid-film force. Both depend on the pressure distribution across the bearing surface. The study of the relationships between pressures and flow is therefore essential to the derivation of design data.

Important considerations include flow rate through restrictors and through bearings, and fluid-film force on bearing surfaces. Fluid properties that govern pressure and steady flow are viscosity and density. For dynamics, the bulk modulus of lubricant elasticity is also important. Additionally, the compressibility of gases requires application of the gas laws.

2.2 Viscosity

Viscosity in a fluid relates shear stress to rate of shear. In Figure 2.1, one surface moves at a speed U parallel to a fixed surface where the two surfaces are separated by a thin film of fluid of thickness h. The friction force F required to shear a surface of area A_f depends on the viscosity according to the relationship:

$$F = \frac{\eta A_f U}{h} \tag{2.1}$$

The shear stress $\tau = F/A_f$ leads to Newton's law of viscosity:

$$\tau = \frac{\eta U}{h} = \eta \frac{du}{dy} \tag{2.2}$$

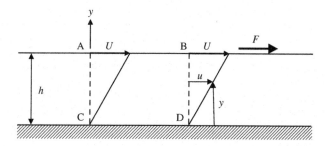

Figure 2.1: Velocity-Induced (Couette) Flow and Viscous Shear Force.

Hydrostatic, Aerostatic and Hybrid Bearing Design.
DOI: 10.1016/B978-0-12-396994-1.00002-4

The viscosity of a Newtonian fluid is constant with shear rate. A simple mineral oil at constant temperature is an example of a Newtonian liquid and air is an example of a Newtonian gas. Polymeric liquids tend to be non-Newtonian because shear stress varies with shear rate due to long chain molecules.

The units of viscosity are *force* × *time/area*. In the Système International, SI units may be quoted as N s/m^2 or Pa s, where 1 N s/m^2 is equivalent to 1000 cP (i.e. 1000 centipoise). In British engineering units, dynamic viscosity is expressed in reyns. The reyn is the name given to the group lbf s/in^2. One reyn is equivalent to 6.895×10^6 cP.

Engineering calculations are invariably based on dynamic viscosity whereas oil companies frequently quote kinematic viscosity, v (i.e. Greek nu), obtained directly from measurements. Kinematic viscosity, v, is related to dynamic viscosity by the oil density, ρ, where

$$v = \frac{\eta}{\rho} \tag{2.3}$$

In British engineering units, kinematic viscosity is expressed in in^2/s. The more common unit, however, is the centistokes (cSt), where 1 cSt is equivalent to 10^{-6} m^2/s and 1.55×10^{-3} in^2/s.

The viscosity of oils decreases with temperature so that it is always necessary to estimate the temperature at which the system will operate. This may require a value to be assumed so that initial design may be performed. This is followed by performing more accurate estimates and repeating the design calculations until sufficient accuracy is achieved.

For hydrostatic and hydrodynamic bearings, it is usual to base the effective viscosity in the bearing on the inlet temperature, T_i, plus a proportion of the temperature rise, ΔT. The simplest estimate of the effective temperature is

$$T_{eff} = T_i + 0.5\Delta T_{max}$$

The term ΔT_{max} is the maximum theoretical temperature rise based on 100% of the heat dissipated by convection into the lubricant. It is often considered (Siew, 1982) to be more accurate in estimating temperature rise, flow rate, and friction to use the proportions:

$$T_{eff} = T_i + 0.8\Delta T_{max}$$

Here, ΔT is based on 60% of the heat dissipated by convection. This method gives almost the same effective temperature on which to base the viscosity but reduces the temperature rise slightly.

If the temperature rise is high, viscosity in the bearing will be different from the viscosity in the control device (Rowe and Stout, 1972). Viscosity is also dependent on pressure, increasing as pressure is raised. The increase in viscosity is not usually significant in hydrostatic bearing calculations, where the majority of applications involve pressures up to 10 MN/m^2 (1470 lbf/in^2). At this pressure, the viscosity of a light mineral oil might be increased by 25%. The designer

Table 2.1: Typical values of dynamic viscosity

Fluid	Temperature (°C)	Dynamic Viscosity	
		(cP)	(μreyn)
Air	18	0.0183	0.002654
Water	20	1.002	0.1453
Water	30	0.7975	0.1156
Machine oil (light)	15.6	113.8	16.5
Machine oil (light)	37.8	34.2	4.959

might well evaluate flow rate on the basis of the viscosity at the average system pressure. This would lead to a minor error in the calculations.

Some typical values of viscosity are given in Table 2.1.

2.3 Density and Consistent Units

The density ρ of a fluid is defined as the mass per unit volume. In order to avoid mistakes in calculation, it is important to use *consistent units* for force, length, mass, and time. Consistent SI units are force (N), length (m), mass (kg), and time (s), so that density is expressed in kg/m^3. Consistent British engineering units are force (lbf), length (ft), mass (slug), and time (s), leading to density expressed in $slugs/ft^3$.

A consistent and more convenient set of British engineering units for lubrication calculations is force (lbf), length (in), mass (lbf in^{-1} s^2), and time (s). This leads to the units of mass density lbf in^{-4}s. Approximate densities of some common materials are listed in Table 2.2.

2.4 Compressibility

Liquids

Liquids are assumed to be incompressible for most calculations. However, at high frequencies, compressibility affects stiffness and damping of a bearing film. Compressibility of a liquid is

Table 2.2: Approximate values of density

Fluid	Temperature (°C)	Density	
		(kg/m^3)	(10^{-6} lbf in^{-4} s)
Air	20	1.208	0.1129
Water	20	1000	93.5
Machine oil (light)	20	875	81.8
Machine oil (light)	100	820	76.67

defined by its "bulk modulus of elasticity", K. The change in volume dV may be related to a change in pressure dp by the equation:

$$dV = -\frac{V}{K} dp \qquad (2.4a)$$

K has the units of pressure. For water at normal temperature and pressure, the bulk modulus $K = 2.068$ GN/m^2 (300,000 lbf/in^2).

Gases

Compressibility of gas affects every aspect of aerostatic design, including geometry and layout of bearing pads, entry recesses, supply pressure, restrictor design, gas flow, power consumption, and stability. Pressure and volume of a gas are related by the general gas law:

$$pv = RT \qquad (2.4b)$$

In the above equation, p is absolute gas pressure, $v = 1/\rho$ is volume per unit mass, ρ is density, T is absolute temperature, and for a perfect gas $R = 8.314$ J/K mol. For a polytropic pressure change:

$$pv^k = \text{a constant} \qquad (2.4c)$$

The exponent k lies in the range $1 < k < \gamma$, where $\gamma = C_p/C_v$ is the ratio of the specific heats of the gas at constant pressure and at constant volume. Change of pressure is accompanied by a change in temperature $T_2/T_1 = (p_2/p_1)^{(k-1)/k}$. For air, $\gamma \approx 1.4$. Conduction of heat into the gas stream quickly restores gas temperature to ambient conditions in most thin-film situations. This allows a small recovery of pressure after gas passes through a jet into a bearing.

2.5 Viscous Flow Between Parallel Plates

Viscous flow is assumed when the length of the flow path is very much longer than the film thickness and when the Reynolds number is less than 2000. This is the condition that usually applies in a bearing pad. If the length of the flow path is less than 20 times the film thickness, the calculation of pressures based on viscous flow may be inaccurate due to pressure drop around the entrance region. An example is radial flow from a recess that is very small. The boundary of the small recess may form an annular orifice restriction so as to reduce inlet pressure. Flow through restrictors is further described in Section 2.7 and Chapter 5. In this section, entrance and exit effects are assumed to be negligible where viscous flow predominates and momentum effects are negligible. In practice, such simplifying assumptions are not always justified.

Pressure-Induced (Poiseuille) Flow

In Figure 2.2, laminar flow takes place between two stationary parallel plates separated by a thin film. Flow is caused by the application of pressure at one end. The flow is resisted by

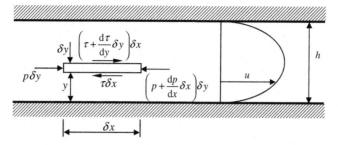

Figure 2.2: Pressure-Induced (Poiseuille) Flow.

shear stresses in the fluid. The velocity at the boundary is zero and increases to a maximum at the center of the thin film. The pressure across the flow is assumed to be constant—that is, $\partial p/\partial y = 0$. The force equilibrium on an element of width z is

$$pz\delta y - \left(p + \frac{dp}{dx}\delta x\right)z\delta y - \tau z\delta x + \left(\tau + \frac{d\tau}{dy}\delta y\right)z\delta x = 0$$

leading to

$$\frac{d\tau}{dy} = \frac{dp}{dx} \tag{2.5}$$

Substituting $\tau = \eta(du/dy)$ leads to

$$\eta\frac{d^2u}{dy^2} = \frac{dp}{dx} \tag{2.6}$$

Integrating twice yields the velocity distribution. The boundary conditions are $du/dy = 0$ and $y = h/2$ and $u = 0$ at $y = 0$ or h:

$$u = \frac{1}{2\eta}\frac{dp}{dx}(y^2 - yh) \tag{2.7}$$

The integral of the velocity is the volumetric flow:

$$q = \frac{zh^3}{12\eta}\frac{dp}{dx} \qquad \text{Volumetric flow}$$

$$m = \frac{\rho zh^3}{12\eta}\frac{dp}{dx} \qquad \text{Mass flow} \tag{2.8}$$

Equation (2.8) for mass flow applies for both liquids and gases. For gas lubrication, further integration to obtain a pressure distribution requires substitution for the variable density term whereas constant density is appropriate for incompressible liquids. For liquid flow through a thin parallel film slot of width z and length l, the pressure gradient $dp/dx = (p_1 - p_2)/l$ is constant, so that

$$q = \frac{zh^3}{12\eta l}(p_1 - p_2) \qquad \text{Liquid flow} \qquad (2.9a)$$

For isothermal gas flow, equation (2.4b) leads to volume flow q_2 at absolute pressure p_2:

$$q_2 = \frac{zh^3}{12\eta l}\frac{(p_1^2 - p_2^2)}{2p_2} \qquad \text{Gas flow} \qquad (2.9b)$$

In most situations, volume flow is stated in terms of free air or gas at ambient pressure. For isothermal conditions, which can be assumed in most cases, flow of free air is increased relative to flow at high pressure. The increased flow is proportional to p_1/p_2.

Velocity-Induced (Couette) Flow

Velocity-induced flow results if one surface moves parallel to the other, as in Figure 2.1. Velocity-induced flow is independent of the pressures. Assuming no slip between the fluid and the surfaces, the velocity is $u = 0$ at a stationary surface and $u = U$ at a moving surface. The velocity gradient is constant so that $du/dy = U/h$. The velocity-induced flow in Figure 2.1 is given by

$$q = \frac{1}{2}Uzh \qquad (2.10)$$

Pressure-Induced Flow Between Circular Parallel Plates

The pressure-induced laminar flow for the circular pad in Figure 2.3 is derived by applying a circular element to equation (2.7), leading to

$$q = \frac{m}{\rho} = \frac{-\pi r h^3}{6\eta}\frac{dp}{dr} \qquad (2.11)$$

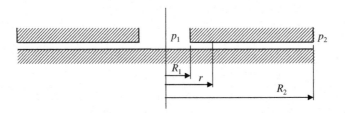

Figure 2.3: Circular Pad.

Mass flow m must be constant at any radius r to ensure continuity. Rearranging equation (2.11) allows the pressure distribution to be expressed in terms of the constant mass flow:

$$dp = -\frac{6\eta q}{\pi h^3}\frac{dr}{r} \quad \text{or} \quad dp = -\frac{6\eta m}{\pi \rho h^3}\frac{dr}{r} \tag{2.12}$$

The pressure distribution for liquid flow is given by a further integration with the boundary condition $p = p_2$ when $r = R_2$:

$$p_1 - p_2 = \frac{6\eta q}{\pi h^3}\cdot\log_e(R_2/R_1) \qquad \text{Liquid} \tag{2.13a}$$

The flow rate for liquid is found from equation (2.13a) by employing the values $p_1 = p_r$ when $r = R_1$ and $p_2 = p_a$ at $r = R_2$ so that $p_1 - p_2 = P_r$, leading to

$$q = P_r\frac{\overline{B}h^3}{\eta} \qquad \text{Liquid} \tag{2.13b}$$

$\overline{B}$ is a flow shape factor and P_r is the gauge value of recess pressure. For circular plates,

$$\overline{B} = \frac{\pi}{6\log_e(R_2/R_1)}$$

For gases, the term for density in equation (2.12) leads to

$$q_2 = \frac{p_1^2 - p_2^2}{2p_2}\frac{\overline{B}h^3}{\eta} \qquad \text{Gas} \tag{2.13c}$$

Flow rate to atmosphere for gas can be expressed as

$$q = P_r\frac{\overline{B}h^3}{\eta}\frac{p_r + p_a}{2p_a} \qquad \text{Gas} \tag{2.13d}$$

where $P_r = p_r - p_a$ is gauge recess pressure. It may be seen that flow of a compressible gas is increased compared with an incompressible fluid by a factor equal to the mean pressure expressed in atmospheres. Flow to atmosphere associated with density ρ_a at pressure p_a is generally termed "flow of free air".

An important conclusion from the liquid and gas versions of equation (2.13) is that flow factor $\overline{B}$ based on pad geometry is the same for both liquid and gas, although the associated pressure terms in equations (2.13) are different.

Combined Pressure- and Velocity-Induced Flow in a Slot

The equilibrium of an element in Figure 2.4 leads to equation (2.6). Integrating with boundary conditions $u = 0$ when $y = 0$ and $u = U$ when $y = h$, the velocity distribution is

$$u = \frac{Uy}{h} - \frac{1}{2\eta}\frac{dp}{dx}(yh - y^2) \tag{2.14}$$

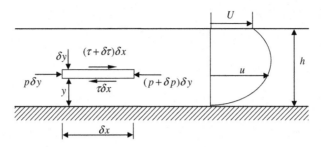

Figure 2.4: Combined Velocity and Pressure-Induced Flows.

Flow is obtained by a further integration:

$$q = \frac{Uhz}{2} - \frac{zh^3}{12\eta}\frac{\mathrm{d}p}{\mathrm{d}x} \tag{2.15}$$

The equation also applies where the surfaces are nonparallel as in an eccentric journal bearing. The integration of equation (2.15) to obtain the pressure distribution for nonparallel surfaces depends on the relationship between h and x and on the boundary conditions for p.

2.6 Combined Pressure- and Velocity-Induced Viscous Flow in a Two-Dimensional Nonparallel Film

In the previous examples, flow has been considered in only one direction, which has been the x direction, defined where possible as the direction of motion. In a real bearing, there is side flow perpendicular to the direction of motion. In a journal bearing, for example, the side flow is the leakage from the bearing. The direction of side leakage is termed the z coordinate.

The Reynolds equation in the following simplified form expresses the relationship between the pressure gradients in the x and z directions and the surface velocity in a two-dimensional nonparallel thin fluid film that can be either compressible or incompressible:

$$\frac{\partial}{\partial x}\left(\frac{\rho h^3}{12\eta}\frac{\partial p}{\partial x}\right) + \frac{\partial}{\partial z}\left(\frac{\rho h^3}{12\eta}\frac{\partial p}{\partial z}\right) = \frac{\rho U}{2}\frac{\mathrm{d}h}{\mathrm{d}x} \tag{2.16}$$

The terms can be identified with the rate of change of mass flows towards a point due to the pressure and velocity effects. In the absence of a source or sink flow, the sum of the terms must be zero at any point between the bearing surfaces. This follows since the volume of fluid is exactly equal to the space available in any zone between the bearing surfaces. This condition of *flow continuity* is embodied in the Reynolds equation. It follows that the Reynolds equation cannot strictly be applied at a cavitation boundary, although this problem is circumvented since the Reynolds equation can be employed to determine the pressure distribution at every point within the continuous film. In a cavitation region, it may be assumed with reasonable accuracy that the pressure is zero.

In theory, the Reynolds equation applies to a continuous film even if the pressures become negative. This situation may seem improbable on first consideration, because it requires the fluid to withstand tensile stresses. However, liquids take time to cavitate and tensile stresses may be measured in journal bearings where the liquid flows through a diverging gap between the bearing surfaces. In practice, the magnitude of tensile stresses is usually small and it is usual to assume that liquid cavitates at zero pressure, leading to reduced load support. Bearing pressures are solved from the Reynolds equation using numerical methods. Analytical solutions can be found for a few idealized cases, but these will be inaccurate for most real bearings.

Bearings employed to position a mass rather than for continuous motion can be designed for zero speed—that is, $U = 0$. This eliminates the term on the right-hand side, so that

$$\frac{\partial}{\partial x}\left(\frac{\rho h^3}{12\eta}\frac{\partial p}{\partial x}\right) + \frac{\partial}{\partial z}\left(\frac{\rho h^3}{12\eta}\frac{\partial p}{\partial z}\right) = 0 \tag{2.17}$$

Equation (2.17) is the form used to obtain most bearing design data for hydrostatic and aerostatic bearings. For hybrid bearings, equation (2.16) is used.

2.7 Flow Through Restrictors

Flow Continuity Between the Restrictor and the Bearing

As explained in Section 2.1, every hydrostatic bearing requires some form of flow control device. The leakage flow through the bearing lands may be calculated for a given recess pressure. The recess pressure may be calculated by applying the relationship between pressure and flow for the restrictor and by also applying the mass-flow continuity condition at this point in the bearing. The situation is illustrated in Figure 2.5 for a single restrictor feeding a circular hydrostatic pad bearing. Mass-flow continuity requires

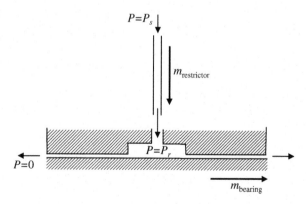

Figure 2.5: Circular Pad Bearing with Restrictor.

$$m_{\text{restrictor}} = m_{\text{bearing}} \qquad (2.18)$$

where mass flow $m = \rho q$. For liquid bearings, density is usually assumed to be constant so that volume flows may be equated, $q_{\text{restrictor}} = q_{\text{bearing}}$. If there are several restrictors in a bearing, the sum of the restrictor flows is equal to the total flow out of the bearing.

Flow Through a Capillary Restrictor

The volume flow through a capillary for laminar isothermal viscous conditions is

$$q = \frac{(P_s - P_r)\pi d_c^4}{128\eta l_c} \qquad \text{Liquid}$$

$$q = \frac{(p_s^2 - p_r^2)}{2p_a} \frac{\pi d_c^4}{128\eta l_c} \qquad \text{Gas flow at atmospheric pressure} \qquad (2.19)$$

The equations apply to tubes where "end-effect" losses are negligible. The ratio l_c/d_c should, if possible, be greater than 100. At values less than 20, significant errors may be involved even at moderate values of Reynolds number. Reynolds number must be less than 2000 for laminar flow to develop. Where velocity v is the average flow velocity and ρ is the fluid density, Re is given by

$$Re = \frac{\rho v d_c}{\eta} \qquad (2.20)$$

Capillary tube restrictors are not usually practicable for aerostatic bearings due to the extremely small bore diameters required. Practicable alternatives are laminar-flow slot restrictors and turbulent-flow orifice restrictors.

Flow Through a Slot Restrictor

A set of slot restrictors for a journal bearing is illustrated in Figure 1.5. By reference to equation (2.9), the rate of laminar isothermal viscous flow through a slot of length l_s, width a_s, and film thickness z_s is

$$q = \frac{(P_s - P_r)a_s z_s^3}{12\eta l_s} \qquad \text{Liquid}$$

$$m = \frac{(p_s^2 - p_r^2)a_s z_s^3}{24\eta RT l_s} \qquad \text{Gas} \qquad (2.21)$$

For isothermal flow, the term RT in equation (2.21) can be replaced by convenient associated values of pressure and density such as $RT_a = p_a/\rho_a$ for flow expressed in terms of the volume of free air. The choice does not affect the mass flow m.

Flow Through an Orifice

A simple orifice is a hole of short length-to-diameter ratio, as illustrated in Figure 2.6. An alternative form of orifice is the annular orifice that depends on the film thickness. An

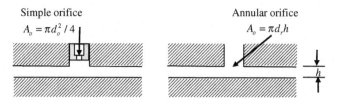

Figure 2.6: A Simple Orifice and an Annular Orifice.

annular orifice has the effect of reducing bearing film stiffness by one-third, as described in Chapter 5.

Liquid Flow

The flow rate for turbulent flow of liquid is given by the following expression, where P_s is the upstream supply pressure and P_r is the downstream pressure from the orifice:

$$q = C_d A_o \sqrt{\frac{2(P_s - P_r)}{\rho}} \qquad \text{Liquid} \qquad (2.22a)$$

C_d is an orifice discharge coefficient that varies with Reynolds number. For liquid flow, typical values are $C_d \approx 0.55$ for $Re = 10^6$, as shown in Figure 2.7. These values also include typical effects of a vena contracta.

Air Flow

For gas flow, C_d varies with Mach number and also with the absolute pressure ratio $K_p = p_r/p_s$. At high values of pressure ratio, values of C_d reduce towards zero. A typical range for C_d is $\approx 0.7-0.96$ for pressure ratios less than 0.7. The upper limit of the range, $C_d = 0.96$, is from experiments by Grewal (1979) for an orifice diameter of 0.3 mm. The lower range is from Powell (1970). Pink and Stout (1978) calibrated a range of orifices of varying diameters for supply pressures in the range $3 < p_s/p_a < 7.8$ for choked flow conditions. Discharge coefficients varied from $C_d = 0.68$ at $d_o = 0.1$ mm to $C_d = 0.88$ at $d_o = 0.3$ mm. This range of values of C_d suggests that it is advisable to calibrate jets for best accuracy.

Gas flow through an orifice is polytropic, leading to a pair of equations (2.22b) and (2.22c):

$$m = C_d A_o \rho_s \sqrt{2 R T_s} \sqrt{\frac{k}{k-1} \left[\left(\frac{p_r}{p_s} \right)^{2/k} - \left(\frac{p_r}{p_s} \right)^{\frac{k+1}{k}} \right]} \qquad \text{Gas flow} \qquad (2.22b)$$

$$m_{\text{choked}} = C_d A_o \rho_s \sqrt{2 R T_s} \sqrt{\frac{k}{k-1} \left[(\overline{p}_c)^{2/k} - (\overline{p}_c)^{\frac{k+1}{k}} \right]} \qquad \text{Gas flow} \qquad (2.22c)$$

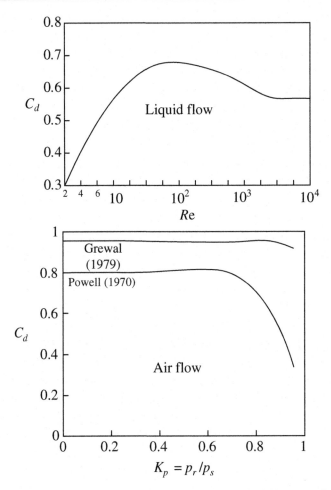

Figure 2.7: Typical Variations of C_d for Orifice Flow.

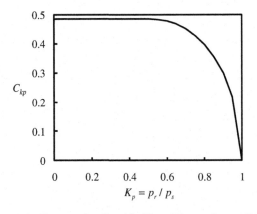

Figure 2.8: Factor C_{kp} for Air Flow Through an Orifice.

The exponent k is approximately equal to C_p/C_v and ρ_s is the gas density at supply pressure p_s and supply temperature T_s. Choked flow occurs when K_p is less than $\bar{p}_c = p_r/p_s = (2/(k+1))^{k/(k-1)}$. The conditions for choked maximum flow are found by differentiating m with respect to K_p and setting to zero. When $p_r/p_s < \bar{p}_c$, it is replaced by $\bar{p}_c$ in equation (2.22b) to give the choked flow rate. The mass-flow rate is then given by $m = C_d A_o \rho_s \sqrt{2RT_s} \cdot C_{kp}$, where the factor C_{kp} applies across the range. Values for air, oxygen, and nitrogen are given in Figure 2.8 for $k = 1.4$.

Further information on the design of orifices is given in Chapter 5, including flow corrections to take account of annular curtain restriction surrounding a simple orifice.

2.8 Recess Pressure and Pressure Ratio

A Capillary-Controlled Bearing

Recess pressure is found by equating flow through the control device and flow through the bearing. This principle is described in detail for a hydrostatic capillary-controlled circular pad, as shown in Figure 2.5.

From equation (2.13), flow through the bearing when $p_1 = p_r$ and $p_2 = 0$ is

$$q_{out} = \frac{P_r \pi h^3}{6\eta \log_e (R_2/R_1)} \qquad \text{Liquid} \qquad (2.23)$$

Equating to restrictor flow:

$$\frac{(P_s - P_r)\pi d_c^4}{128\eta l_c} = \frac{P_r \pi h^3}{6\eta \log_e (R_2/R_1)}$$

Rearranging terms, recess pressure is

$$P_r = \frac{P_s \dfrac{d_c^4}{128 l_c}}{\dfrac{d_c^4}{128 l_c} + \dfrac{h^3}{6\log_e (R_2/R_1)}} \qquad (2.24)$$

A general form for a single pad with a laminar flow restrictor is

$$P_r = \frac{P_s}{1 + \bar{B}K_c h^3} \qquad \text{Liquid} \qquad (2.25a)$$

For gas flow, the absolute pressure is given by

$$p_r^2 = \frac{p_s^2 + p_a^2 \bar{B}K_c h^3}{1 + \bar{B}K_c h^3} \qquad \text{Gas} \qquad (2.25b)$$

Gauge recess pressure $P_r = p_r - p_a$ is

$$P_r = \sqrt{\frac{p_s^2 - p_a^2}{1 + \overline{B}K_c h^3}} \qquad \text{Gas} \qquad (2.25c)$$

The flow factor for a circular pad is

$$\overline{B} = \frac{\pi}{6\log_e(R_2/R_1)}$$

and the restrictor factor for a capillary is

$$K_c = \frac{128 l_c}{\pi d_c^4}$$

It may be noted that if the viscosity of the lubricant as it passes through the restrictor is the same value as the viscosity in the bearing the recess pressure will be independent of viscosity, as in equation (2.25). This is an advantage of capillary control as it makes the bearing relatively insensitive to changes in average temperature of the lubricant. As a note of caution, however, if the flow rate is too low for a bearing lubricated with mineral oil running at high speed, the temperature will rise as the lubricant passes through the bearing. Temperature rise leads to a difference between the viscosity of the lubricant in the restrictor and the viscosity of the lubricant in the bearing. In extreme cases, this causes a loss of performance. Matching pumping power to friction power prevents this potential problem, as explained in later chapters, where it is shown that power ratio is an important design criterion for bearings that operate at speed.

An Orifice-Controlled Bearing

Equating liquid flow through an orifice from equation (2.22a) to flow through a bearing from equation (2.23) leads to P_r. The solution is

$$P_r = \frac{\sqrt{1 + 4P_s K_o \left[\frac{\overline{B}h^3}{\eta}\right]^2} - 1}{2K_o \left[\frac{\overline{B}h^3}{\eta}\right]^2} \qquad \text{Liquid} \qquad (2.26)$$

The orifice coefficient is $K_o = \rho/[2(C_d A_o)^2]$, where A_o is the restriction area of the orifice restrictor. For a circular pad, the flow shape coefficient is

$$\overline{B} = \frac{\pi}{6\log_e(R_2/R_1)}$$

The flow coefficient is independent of the type of restrictor. Equation (2.26) is a general form for orifice-controlled pads. Values of $\overline{B}$ for pads of various shapes are given in Chapter 4.

Recess pressure depends on viscosity in an orifice-controlled bearing. This is important because, as the system warms up during running, viscosity is reduced, causing reduced bearing film thickness. Capillary or slot-controlled bearings are relatively insensitive to changes in viscosity.

Equivalent expressions for aerostatic bearings are given below in Section 2.11.

Pressure Ratio and the Design Condition

Gauge pressure ratio P_r/P_s is given by equations such as (2.25) and (2.26). More generally, values are given by design charts such as those in Chapters 7 and 8 for variations with load and displacement. Pressure ratio varies with film thickness and applied load. If one value of the film thickness is denoted the design condition—that is, h_o—the particular value of pressure ratio is termed β, which is the ratio P_{ro}/P_s, where P_{ro} is the value of pressure P_r when $h = h_o$. For liquid hydrostatic bearings, ambient pressure is usually taken as zero and all pressures are relative to ambient conditions so that gauge pressure ratio is

$$\beta = \frac{P_{ro}}{P_s} \qquad \text{Liquid}$$

The usefulness of designating a design condition is that it allows load, flow, and stiffness variations to be explored for different values of β. A value $\beta = 0.5$ allows loads to be increased or reduced equally in either sense.

Traditionally, the symbol for gauge pressure ratio K_{go} is employed for gas:

$$K_{go} = \frac{p_{ro} - p_a}{p_s - p_a} \qquad \text{Gas}$$

2.9 Bearing Load

The force exerted by a liquid film against two bearing members must for equilibrium balance the externally applied load. Bearing load is determined by summing *pressure × area* across elements of the bearing surface, as illustrated in Figures 1.1 and 1.2 for two idealized situations. The following examples illustrate the process applied to practical bearing shapes.

A Plane Bearing Pad

The shape of the pressure distribution under a circular bearing is found by integrating equation (2.12) across the bearing lands shown in Figure 2.3 from R_1 to r. With an inlet pressure p_r this gives

$$p - p_a = (p_r - p_a)\left[1 - \frac{\ln(r/R_1)}{\ln(R_2/R_1)}\right] \qquad \text{Liquid}$$

$$p^2 - p_a^2 = (p_r^2 - p_a^2)\left[1 - \frac{\ln(r/R_1)}{\ln(R_2/R_1)}\right] \qquad \text{Gas}$$

(2.27)

A typical effect of compressibility on the resulting pressures is shown in Figure 2.9. Pressures are increased by compressibility at higher pressures so that an aerostatic bearing carries increased load.

Bearing load is supported due to both the recess pressure and the land pressures, so that

$$W = (p_r - p_a)\pi R_1^2 + \int_{R_1}^{R_2} (p - p_a)(2\pi r)dr \tag{2.28}$$

The expression for W may be integrated for liquid flow, leading to

$$W = \frac{\pi P_r(R_2^2 - R_1^2)}{2\log_e(R_2/R_1)} \tag{2.29}$$

This may be expressed in alternative forms according to convenience:

$$W = P_r\pi R_2^2\overline{A} \quad \text{or} \quad W = P_r A\overline{A} \quad \text{or} \quad W = P_r A_e \tag{2.30}$$

The area of the bearing surface is A, and $\overline{A}$ is the fraction by which the area must be multiplied to determine the effective area A_e.

Equations (2.29) and (2.30) may also be used as an approximate guide for gas bearings employing gauge pressure $P_r = p_r - p_a$. For the example in Figure 2.9, $R_2/R_1 = 10$ and it is found that $\overline{A}_{hs} = 0.215$, whereas $\overline{A}_{ae} = 0.293$, a 36% increase in load supported. In practice, a hydrostatic pad usually has a larger recess than an aerostatic pad, so that a hydrostatic pad is likely to carry a larger load for the same recess pressure. The increase in $\overline{A}$ for other pad shapes is almost always smaller than 36%. The exception is where an even smaller recess is employed for a circular pad with central admission. In this case, $\overline{A}$ is reduced as R_2/R_1 is increased and the differences between gas and liquid loads supported become larger. This situation is illustrated in Figure 4.1.

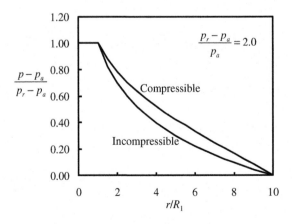

Figure 2.9: Effect of Compressibility on Circular Pad Pressures.

Load support for hydrostatic bearings is usually a conservative guide for aerostatic bearings of the same geometry. Sometimes, orifice-fed virtual recesses for aerostatic bearings replace equivalent real recesses for hydrostatic bearings. Orifice-fed virtual recesses suffer dispersion losses as described in Chapters 4 and 5, leading to a reduction of load support.

An Opposed-Pad Plane Bearing

An opposed-pad bearing (Figure 2.10) is used to withstand applied loads acting in either direction or to increase bearing stiffness without increasing the weight carried. In this case, bearing load depends on the difference between the two pad thrusts:

$$W = P_2 A_{e2} - P_1 A_{e1} \tag{2.31}$$

Journal Bearings

Figure 1.3 shows the circumferential pressure distribution in a four-recess journal bearing. The pressure due to one recess is represented in Figure 2.11. For accurate load integration, the pressures must be summed separately for two independent directions, such as in the horizontal and

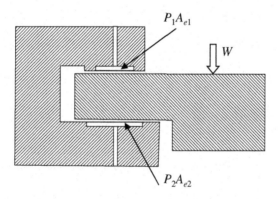

Figure 2.10: Opposed-Pad Bearing.

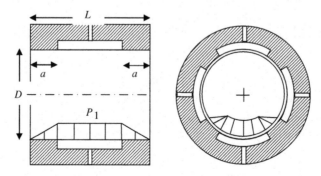

Figure 2.11: Four-Recess Journal Bearing: Pressure Due to One Recess.

vertical directions. This process may be carried out numerically according to equations of the form:

$$W_v = \int\limits_{0}^{2\pi} \int\limits_{0}^{L} P \frac{D}{2} \cos\theta \; d\theta \; dz$$

$$W_h = \int\limits_{0}^{2\pi} \int\limits_{0}^{L} P \frac{D}{2} \sin\theta \; d\theta \; dz$$

(2.32)

where W_v and W_h are the vertical and horizontal components of the load.

Approximate methods are sometimes employed when a quick answer is required and data are not available. The effect of each recess is estimated and the contributions from each added up. The basis may be seen from Figure 2.11. The contribution to the vertical load due to P_1 acting on the projected area $(L - a)D\sin(\pi/4)$ is

$$W_1 = P_1(L - a)D \sin(\pi/4)$$

Adding together the contributions from each recess:

$$W_v = (P_1 - P_3)(L - a)D \sin(\pi/4)$$

$$W_h = (P_2 - P_4)(L - a)D \sin(\pi/4)$$

2.10 Use of Normalized Data

Data do not need to be derived from first principles except in unusual cases. A small amount of normalized data serves a variety of bearing shapes and sizes. For example, equation (2.30) for a single pad is

$$\text{Load} = \text{recess pressure} \times \text{pad area} \times \text{area factor } \overline{A}$$

The dimensionless group $W/P_r A$ is said to be "normalized". A normalized value depends only on bearing shape and is given the symbol $\overline{A}$. A single normalized value $\overline{A}$ covers all combinations of W with P_r and A for a particular land-width ratio.

Convenient dimensionless groups consist of properties chosen by the designer. In this example, recess pressure P_r is required whereas supply pressure P_s is a more convenient choice. Another group that takes account of variations in pressure ratio is $\overline{W} = W/P_s A_e$. For a single pad, $\overline{W} = P_r/P_s$. For an opposed pad bearing, $\overline{W} = (P_1 - P_2)/P_s$.

In the following chapters, normalized data are used as convenient for a particular application. The use of normalized data has another advantage that any set of consistent units can be employed when evaluating data. Some common groups are given in Table 2.3.

Table 2.3: Common design groups

Property	Design Group	Symbol
Concentric pressure ratio	$\dfrac{P_{ro}}{P_s}$	β for liquids
Concentric gauge pressure ratio	$\dfrac{P_{ro}}{P_s} = \dfrac{p_{ro} - p_a}{p_s - p_a}$	K_{go} for gases
Effective area	$\dfrac{A_e}{A}$	$\overline{A}$
Load support	$\dfrac{W}{P_s A_e}$	$\overline{W}$
Flow shape factor (liquid flow)	$\dfrac{q\eta}{P_r h^3}$	$\overline{B}$
Flow shape factor (gas flow)	$\dfrac{2q\eta p_a}{(p_r^2 - p_a^2)h^3}$	$\overline{B}$
Flow pressure factor	$\dfrac{q\eta}{P_s \overline{B} h_o^3}$	$\overline{Q}$
Speed parameter (journals)	$\dfrac{\eta N}{P_s}\left(\dfrac{D}{2h_o}\right)^2$	S_h
Speed parameter (pads)	$\dfrac{\mu L V}{P_s h_o^2}$	S_h
Film stiffness	$\dfrac{\lambda h_o}{P_s A_e}$	$\overline{\lambda}$

Taking a normalized data value such as $\overline{A}$ from a data chart, it is possible to choose all the parameters within the group except one, which can then be evaluated. For example, a designer might look up values of $\overline{A}$ and $\overline{W}$ from data charts for a particular shape of bearing such as a rectangular multi-recess pad. Using a value of applied load W to be supported by a bearing, the designer freely chooses A before evaluating a suitable value of P_s. Alternatively, the designer can freely choose P_s before evaluating A:

$$P_s = \frac{W}{\overline{W}A_e} = \frac{W}{\overline{W}A\overline{A}} \tag{2.33}$$

■ *Example 2.1 Use of Normalized Data*

For a single-recess rectangular plane pad, the area shape factor is found from a design chart to be $\overline{A} = 0.65$. The load factor from a design chart for a single pad is $\overline{W} = 0.5$. In other words, the recess pressure is to be half the supply pressure at the normal operating design load. The applied load to be supported is $W = 1200$ N. The designer wishes to contain the pad within a total area $A = 800$ mm^2. Employing consistent SI units, the area is 800×10^{-6} m^2. The supply pressure required is given by equation (2.33),

$$P_s = \frac{W}{\overline{W}A\overline{A}} = \frac{1200}{0.5 \times 800 \times 10^{-6} \times 0.65} = 4.62 \times 10^6 \text{ N/m}^2 (670 \text{ lbf/in}^2)$$

■

2.11 Aerostatic Bearings—Summary of Relationships

Shape factors $\bar{A}$ and $\bar{B}$, used for hydrostatic design, are also used for design of aerostatic bearings. This is demonstrated below for absolute film pressures up to approximately 4 bar, equivalent to a gauge pressure of 3 bar (45 lbf/in^2 gauge). For high pressures and large film thickness, inertia effects at the entrance region to the pad cause loss of load support for orifice-fed radial flow bearings. A prototype aerostatic bearing should therefore be tested experimentally prior to finalization of a design. This is particularly relevant for critical applications. While orifice-fed aerostatic bearings are widely applied there is something to be said for viscous flow line-source restrictors. If orifice restrictors are employed, it is sometimes recommended that bearings should be designed to avoid supersonic flow.

Alternative sources of data for air and other bearings are listed in the Further Reading at the end of the chapter.

The Isothermal Assumption

Conditions are usually assumed to be isothermal. This is reasonable because conduction into and from a bearing is usually sufficient to maintain almost constant temperature. In situations where the pressure suffers a sudden drop, as in the case of flow through an orifice, expansion of the gas causes a drop in temperature and is usually assumed adiabatic for calculation of downstream pressures. However, temperature changes are usually small in thin films and conduction quickly restores ambient conditions.

Volumetric Flow of Free Air

Assuming temperatures quickly return to ambient, the term $RT = p_2/\rho_2 = p_a/\rho_a$, where for air exhausting to atmosphere, $\rho_a = 1.208$ kg/m^3. For isothermal conditions, RT can therefore be eliminated from flow equations. For example, equation (2.21) for flow through a slot restrictor becomes

$$q == \frac{a_s z_s^3}{12\eta y_s} \frac{(p_s^2 - p_r^2)}{2p_a}$$

Since

$$\frac{p_1^2 - p_a^2}{2p_a} = (p_1 - p_a)\frac{(p_1 + p_a)}{2p_a}$$

it is seen that isothermal flow through a gas bearing is found by multiplying the incompressible flow by a factor $f_m = (p_1 + p_a)/2p_a$, which is proportional to the mean pad pressure.

This factor gives volumetric flow of free air at the pressure p_a, as shown by the example below.

The relationships for aerostatic flow are summarized below.

Gas Laws

$$\frac{p}{\rho} = RT \qquad \text{Universal gas law}$$

$$\frac{p_1}{\rho_1} = \frac{p_2}{\rho_2} = RT \qquad \text{Isothermal gas law} \tag{2.34}$$

$$\frac{p_1}{\rho_1^k} = \frac{p_2}{\rho_2^k} = RT \qquad \text{Polytropic gas law}$$

Gas Flow Through a Slot Restrictor

For gas flow through a slot restrictor, equation (2.21) has the form:

$$m = \rho q = \frac{\rho_a a_s z_s^3}{12\eta RT_s y_s} \frac{(p_s^2 - p_r^2)}{2p_a} \tag{2.35}$$

Gas Flow Through a Circular Pad

$$m = \frac{\rho_a \pi h^3}{6\eta \log_e (R_2/R_1)} \frac{(p_r^2 - p_a^2)}{2p_a} \tag{2.36}$$

Gas Flow Through a Capillary Restrictor

$$m = \frac{(p_s^2 - p_r^2)}{2p_a} \cdot \rho_a \cdot \frac{\pi d_c^4}{128\eta l_c} \tag{2.37}$$

Gas Flow Through an Orifice Restrictor

Air pressure drops adiabatically from the supply pressure p_s to the downstream pressure p_r:

$$m = C_d A_o \rho_s \sqrt{2RT_s} \sqrt{\frac{k}{k-1} \left[\left(\frac{p_r}{p_s}\right)^{2/k} - \left(\frac{p_r}{p_s}\right)^{\frac{k+1}{k}} \right]} \qquad \text{Unchoked}$$

$$\tag{2.38}$$

$$m_{\text{choked}} = C_d A_o \rho_s \sqrt{2RT_s} \sqrt{\frac{k}{k-1} \left[(\bar{p}_c)^{2/k} - (\bar{p}_c)^{\frac{k+1}{k}} \right]} \qquad \text{Choked}$$

The condition for choked flow is that $p_r/p_s \leq \bar{p}_c = (2/(k+1))^{k/(k-1)}$. For air, $k = 1.4$ and $\bar{p}_c = 0.528$.

The value of C_d varies with the pressure ratio $K_p = p_r/p_s$. For pressure ratios K_p less than 0.7, $C_d \approx 0.7-0.96$. For high values of pressure ratio, C_d drops rapidly towards zero. The flow reaches a maximum when the pressure ratio has a value 0.528 for air and remains constant for lower values. To achieve the high value of $C_d = 0.96$, care is necessary with the nozzle design, with particular attention to the inlet angle to the nozzle and the smoothness of the orifice, and also to the accuracy and smoothness of the inlet and exit conditions. In most cases and particularly for small jets, C_d will be lower, as explained in Section 2.7. The simplest procedure for satisfactory results is to calibrate jets under the required pressure conditions using a flow meter.

Choked jets are usually avoided for aerostatic bearings to reduce the risk of pneumatic hammer, a form of instability that sometimes occurs in orifice-fed aerostatic bearings. The risk of pneumatic hammer is further discussed in Chapter 5.

With temperature recovery, free air flow is $q_a = m/\rho_a$ and $RT_s = p_a/\rho_a$. These terms are employed to simplify application of equations (2.38) in Chapter 5 and in worked examples.

The application of orifice control for aerostatic bearings is further discussed in Chapter 5.

Recess Pressure in a Capillary-Controlled Pad

$$\left(\frac{p_r}{p_s}\right)^2 = \frac{1 + \left(\frac{p_a}{p_s}\right)^2 \bar{B}K_c h^3}{1 + \bar{B}K_c h^3} \tag{2.39}$$

Recess Pressure in an Orifice-Controlled Pad

$$\left(\frac{p_r}{p_s}\right)^2 = \left(\frac{p_a}{p_s}\right)^2 + B' \cdot \left(\frac{p_a}{p_s}\right)\left(\frac{p_r}{p_s}\right)^{\frac{1}{k}}\sqrt{1 - \left(\frac{p_r}{p_s}\right)^{\frac{k-1}{k}}} \qquad \text{Unchoked flow}$$

$$\left(\frac{p_r}{p_s}\right)^2 = \left(\frac{p_a}{p_s}\right)^2 + B' \cdot \left(\frac{p_a}{p_s}\right)(\bar{p}_c)^{\frac{1}{k}}\sqrt{1 - (\bar{p}_c)^{\frac{k-1}{k}}} \qquad \text{Choked flow} \tag{2.40}$$

where

$$B' = \frac{2\eta C_d A_o}{p_a \bar{B} h^3}\sqrt{\frac{2k}{k-1}RT}$$

Virtual Recesses

Later chapters will make reference to virtual recesses for design of bearing pads. A virtual recess is where an area of bearing land is surrounded by a ring of orifices or entry grooves to create the effect of a constant area of pressure, as in a real recess. Virtual recesses are employed instead of actual recesses to improve the speed of response of aerostatic bearings and improve dynamic performance. If the jets or entry grooves surround 70% of the perimeter of the land area, there will be some loss of load support and some reduction of flow, as explained in Section 4.1.

■ *Example 2.2 Estimation of Flow and Load for a Circular Aerostatic Pad*

For a circular plane pad having a small central recess, the flow factor for the chosen shape is $\overline{B} = 0.227$, where $R_2/R_1 = 10$. The load factor is $\overline{A} = 0.215$. The gas pressures are $p_a = 1$ bar and $p_r = 3$ bar. Evaluate flow and approximate load support for an outside radius $R_2 = 1.5$ cm and a bearing film thickness 10 μm. The ambient density of air at room temperature is 1.21 kg/m^3. The viscosity of air at room temperature is taken as 0.0182 cP.

Solution

The mean pressure is

$$p_r = 3 \times 0.1$$

$$= 0.3 \text{ MN/m}^2 \text{ (44.1 lbf/in}^2 \text{ absolute pressure or 29.4 lbf/in}^2 \text{ gauge pressure)}$$

$$p_a = 0.1 \text{ MN/m}^2$$

$$\rho_a = 1.21 \text{ kg/m}^3$$

$$f_m = \frac{p_r + p_a}{2p_a} = \frac{0.3 + 0.1}{2 \times 0.1} = 2$$

$$q = f_m \cdot \frac{(p_r - p_a)\overline{B}h_o^3}{\eta} = 2 \times \frac{(0.3 - 0.1) \times 10^6 \times 0.227 \times (10^3 \times 10^{-18})}{0.0000182}$$

$$= 5.02 \times 10^{-6} \text{ m}^3/\text{s} \text{ (3.0 l/min)}$$

$$m = \rho_a q = 1.21 \times 5.02 \times 10^{-6} = 6.07 \text{ kg/s}$$

The total pad area is $A = \pi R_2^2 = \pi \times 0.015^2 = 0.000707 \text{ m}^2$

Load support is $W = (p_r - p_a)A\overline{A} = (0.3 - 0.1) \times 10^6 \times 0.000707 \times 0.215 = 30.4 \text{ N}$

■

References

Grewal, S. S. (1979). *An investigation of externally-pressurised orifice-compensated air journal bearings with particular reference to misalignment and inter-orifice variations.* CNAA PhD thesis. Liverpool Polytechnic.

Pink, E. G., & Stout, K. J. (1978). *A comparison of the performance of orifice compensated and slot entry gas lubricated journal bearings.* Internal report. Leicester Polytechnic.

Powell, J. W. (1970). *The Design of Aerostatic Bearings.* The Machinery Publishing Company.

Rowe, W. B., & Stout, K. J. (1972). Viscosity variations in hydrostatic bearings. *Tribology International, 5*(4; Dec.), 262–264.

Siew, A. H. (1982). *A computational approach to the design and performance optimization of solid and porous journal bearings.* PhD thesis. Cranfield Institute of Technology.

Further Reading

Gross, W. A. (1962). *Gas Film Lubrication.* Wiley.

Neale, M. J. (1973). *Tribology Handbook.* London: Butterworths.

Prandtl, L., & Tietjens, O. G. (1957). *Fundamentals of Hydro- and Aeromechanics.* New York: Dover Publications.

Schlichting, H. (1967). *Boundary Layer Theory* (6th edn.). New York: McGraw-Hill.

Streeter, V. L. (1962). *Fluid Mechanics* (3rd edn.). New York: McGraw-Hill.

Tabor, D. (1979). *Gases, Liquids and Solids.* Cambridge University Press.

Wilcock, D. F., & Booser, E. R. (1957). *Bearing Design and Application.* New York: McGraw-Hill.

CHAPTER 3

Power, Temperature Rise, and Shape Optimization

Summary of Key Design Formulae

$$H_p = P_s q \qquad \text{Hydrostatic pumping power}$$

$$H_p = p_a q_a \log_e\left(\frac{p_s}{p_a}\right) \qquad \text{Aerostatic pumping power}$$

$$H_f = \frac{\eta A_f U^2}{h} \qquad \text{Friction power}$$

$$K = H_f / H_p \qquad \text{Power ratio}$$

$$\Delta T = \frac{P_s(1 + K)}{\rho c} \qquad \text{Temperature rise per pass}$$

$$A_f = A_t - \frac{3}{4}A_r \qquad \text{Friction area of recessed bearing}$$

3.1 Introduction

For low-speed bearings, the best bearing shape, clearance, and pressure ratio yield maximum load support for minimum power consumption.

For high-speed bearings, it is important to avoid high temperatures and excessive cavitation of the lubricating film leading to a range of problems. Problems are lessened by design for *minimum power*. The technique described is simple and protects against "hot spots". The design technique simplifies selection of supply pressure, viscosity, film thickness, and bearing area.

3.2 Pumping Power H_p

Pumping power is the power to pump fluid through a bearing. For hydrostatic bearings,

$$H_p = P_s q \qquad \text{Hydrostatic} \tag{3.1}$$

Additional power is required to compress gas. For aerostatic bearings,

$$H_p = p_a q_a \log_e\left(\frac{p_s}{p_a}\right) \qquad \text{Aerostatic} \tag{3.2}$$

Hydrostatic, Aerostatic and Hybrid Bearing Design.
DOI: 10.1016/B978-0-12-396994-1.00003-6

Pumping power is therefore greater for aerostatic bearings. Pumping power is doubled for the same mass flow with $\beta = K_{go} = 0.5$ and $P_s/p_a = 4$.

3.3 Friction Power H_f

Friction power is the power to move the bearing. Friction power is the product of sliding friction force F and speed U. The friction force between the bearing surfaces is $F = \eta A_f U/h$, where A_f is the friction area.

$$H_f = \frac{\eta A_f U^2}{h} \tag{3.3}$$

Friction area in recessed bearings depends on recess area A_r and land area A_l. Friction power also depends on bearing film thickness and recess depth. It is customary to make recess depth h_r large compared to film thickness h at the bearing lands. This allows recess pressure to spread over a recess and increases load support. It also reduces recess friction, which is not usually negligible. In practice, recirculation flow takes place in the recesses and tends to increase recess friction. This effect was analyzed by Shinkle and Hornung (1965). An approximate expression for friction power taking account of recirculation is

$$H_f = \eta U^2 \left(\frac{A_l}{h} + \frac{4A_r}{h_r} \right) \tag{3.4}$$

Recess depth for hydrostatic bearings is usually several times larger than the bearing film thickness in order to reduce pressure variations across the recess. On this basis, a generous estimate of friction area in recessed bearings is

$$A_f = A - \frac{3}{4}A_r \tag{3.5}$$

Friction area for plain nonrecessed bearings is given by $A_f = A$, where A is the total area of the sliding contact.

For aerostatic bearings, the same plain bearing calculation is employed (Tawfik et al., 1981). Friction power for aerostatic bearings is very low compared with pumping power, except for very-high-speed bearings. Friction power is often ignored for low-speed aerostatic bearings.

3.4 Power Ratio K

Heat generated in the bearing is equal to the total power. Total power is friction power plus pumping power:

$$H_t = H_p + H_f \tag{3.6}$$

An alternative expression for total power is

$$H_t = H_p(1 + K) \tag{3.7}$$

where power ratio is $K = H_f/H_p$. It follows that temperature rise is proportional to K.

Since power ratio has an easily remembered value such as 0 or 1 for an optimum bearing, it is the key to good bearing design. In a zero-speed bearing $K = 0$: the bearing has purely hydrostatic or aerostatic load support. At speed, with $K = 1$, hydrostatic and hydrodynamic support loads are of the same order and power is in the range for minimum power. The conclusion for aerostatic bearings is similar.

Any bearing where $K \geq 1$ is designated as "high speed" since hydrodynamic effects are significant. Any bearing where $K < 1$ is designated as "low speed" and speed effects are ignored. Most aerostatic bearings are "low speed" since aerodynamic load support is usually negligible.

Power ratio K can be made high or low according to the viscosity and film thickness selected. Summarizing, suitable ranges for K are:

$K = 0$ to 1:   Purely hydrostatic or aerostatic load support at low or moderate speeds
$K = 1$ to 3:   High-speed optimized hybrid hydrostatic or aerostatic bearings
$K = 3$ to 9:   High-speed hybrid plain hydrostatic or aerostatic bearings for higher loads.

3.5 Temperature Rise ΔT

Oil is a relatively poor conductor of heat when compared with water or metals. For many applications, but particularly for small high-speed bearings with large flow, energy is almost wholly convected from the bearing by the oil and is then cooled elsewhere in the circuit. Heat is usually dissipated within a hydraulic power unit by means of a water cooler, air radiator, or by refrigeration. In large slow-moving water-lubricated machines, a larger proportion of the heat is dissipated by conduction through the machine structure. In such equipment, temperature rise is less of a problem. It is therefore usual in bearing design to estimate temperature rise of the oil based on pure convection.

Temperature rise for a single pass of lubricant is the maximum temperature difference between entry and exit from the bearing. It is not the same as rise in bulk temperature, which is estimated from the cooling capability of the hydraulic circuit.

Power H_t is converted into heat within the lubricant according to

$$\text{heat} = \text{mass flow} \times \text{specific heat capacity} \times \text{temperature rise}$$
$$H_t = \rho q c \Delta T \tag{3.8}$$

where c is the specific heat capacity of the lubricant. It is unnecessary to include a mechanical equivalent of heat J using consistent units since mechanical energy and heat energy expressed in consistent units are equivalent, for example 1 Nm $\equiv$ 1 J.

From equations (3.7) and (3.8), temperature rise per pass of the lubricant is

$$\Delta T = \frac{H_p(1+K)}{q\rho c}$$

And canceling flow terms in the numerator and denominator, temperature rise simplifies to

$$\Delta T = \frac{P_s(1+K)}{\rho c} \tag{3.9}$$

In other words, temperature rise depends mainly on supply pressure and power ratio.

Low-Speed Temperature Rise

Writing $K = 0$ to define a low-speed bearing, maximum temperature rise for a hydrostatic bearing is

$$\Delta T = \frac{P_s}{\rho c} \tag{3.10}$$

For aerostatic bearings, expansion of a gas causes a temperature drop except at high power ratios, when viscous losses are increased.

For light machine oil, typical properties are:

Specific heat capacity (constant pressure) $c = 2120$ J/kg K
Density $\rho = 855$ kg/m^3
Gauge supply pressure $P_s = 1$ MN/m^2

Inserting these values in (3.10):

$$\Delta T = \frac{10^6}{2120 \times 855} = 0.55\,°C$$

For other pressures:

$$\Delta T = 0.55 \times 10^{-6} \times P_s \qquad (\text{SI units}\,°C)$$
$$\Delta T = 0.007 \times P \qquad (\text{ips units}\,°F)$$

High-Speed Temperature Rise

For light machine oil, maximum temperature rise is increased according to the power ratio:

$$K = 1: \qquad \Delta T = 1.1 \times 10^{-6}P_s \qquad (\text{SI units}\,°C)$$
$$\Delta T = 0.014 \times P_s \qquad (\text{ips units}\,°F)$$
$$K = 3: \qquad \Delta T = 2.2 \times 10^{-6}P_s \qquad (\text{SI units}\,°C)$$
$$\Delta T = 0.028 \times P_s \qquad (\text{ips units}\,°F)$$

3.6 Minimum Power as an Optimization Criterion

There are a large number of variables in bearing design and a designer has to select a combination, each variable having the best value.

For simplicity, a single criterion is recommended: *maximum load for minimum power.* A designer can vary from optimal, if desired, but optimum is a good place to start.

3.7 Minimum Power for Low-Speed Bearings (K = 0)

For low-speed bearings where $K = 0$, total power is equal to pumping power, $H_t = H_p$. At low speeds, bearing geometry and land widths for hydrostatic bearings may be optimized as follows.

The ratio of power to load support is $H_p/W = P_s q/W$. By reference to Section 2.9, this can be expressed in terms of the design variables:

$$\frac{H_p}{W} = \frac{\beta W\, h^3}{A^2}\frac{\overline{B}}{\eta\, \overline{A}^2} \tag{3.11}$$

Extracting shape dependent variables:

$$\frac{\overline{B}}{\overline{A}^2} = \overline{H}_p \tag{3.12}$$

Maximum load for minimum power is achieved for the land width that minimizes $\overline{H}_p$. An exact value is not critical near an optimum, so that the following recommendations arise from computations of $\overline{H}_p$ for various bearing shapes. The conclusion for aerostatic bearings is similar.

Optimum Land-Width Ratio a/L for Low-Speed Bearings

Hydrostatic Bearings

Optimum land-width ratio for a rectangular pad is approximately

$$a = 0.25 \times L$$

where L is the shorter of bearing width B and length L. The land-width ratio is generally termed *a/L*, although different symbols are sometimes used if more convenient.

The recommendation is demonstrated by the following examples:

1. For a *rectangular pad*, 80 × 100 mm, optimum land width is $a \approx 0.25 \times 80 = 20$ mm.
2. For a *journal bearing*, 150 mm in diameter and 100 mm long, optimum axial land width is $a \approx 0.25 \times 100 = 25$ mm.

3. For a *circular pad*, 50 mm in diameter with central lubricant admission, optimum land width is $a \approx 0.25 \times 50 = 12.5$ mm.
4. For a *circular annular thrust pad* (Figure 1.8), 160 mm outside diameter and 80 mm inside diameter with annular admission, land width should be $a \approx 0.25 \times (80 - 40) = 10$ mm.
5. For a *multi-recess annular thrust pad*, the inter-recess land width would usually be 1–2 times this value, i.e. 10–20 mm.

Aerostatic Bearings

The above recommendations for land-width ratio are also approximately valid for aerostatic bearings. However, there is an important further requirement for aerostatic bearings to reduce the risk of pneumatic hammer instability. For aerostatic bearings, it is generally recommended that recesses should be avoided or of very small volume. Enclosing a central land area with a row of holes or slots provides a high-pressure region or a "virtual recess". A virtual recess greatly reduces the volume of gas in a bearing compared with an actual recess, although consideration should be given to the introduction of dispersion losses, as discussed in the next two chapters.

Optimum Gauge Pressure Ratio $\beta = K_{go}$ for Low-Speed Bearings

Gauge pressure ratio is the ratio P_{ro}/P_s at a bearing film thickness h_o. Pressure ratio is usually chosen for maximum load range or maximum film stiffness at the design value of film thickness h_o. The pressure ratio that achieves maximum load range for most bearings is

$$\beta = K_{go} = 0.5$$

Further information on pressure ratio for particular applications is given in later chapters.

Optimum Clearance h_o for Low-Speed Bearings

Clearance is chosen to be a minimum, consistent with tolerances for machining, thermal expansion, and structural deflection. Selection of clearances is discussed further in Chapter 6.

3.8 Minimum Power for High-Speed Recessed Bearings

Optimum Land-Width Ratio a/L for High-Speed Bearings

Recommended land width is unaltered for a high-speed recessed bearing. This may be seen by substituting $H_t = (1 + K)H_p$ for H_p in equation (3.12). For any value of K, it is still necessary to minimize $\overline{H}_p$ and hence optimum land width is $0.25 \times L$ as before.

Optimum Viscosity η for High-Speed Bearings

The designer may minimize total power by choosing optimum viscosity as follows. Viscosity is varied with other parameters fixed. For example, supply pressure is fixed to support the applied load and clearance is a minimum consistent with tolerances. Total power of a hydrostatic bearing in terms of basic variables is then

$$H_t = \frac{P_s^2 \beta \overline{B} h^3}{\eta} + \frac{\eta A_f U^2}{h} \tag{3.13}$$

The optimum viscosity η is found by calculating values of total power from this expression until a minimum is found. The optimum is also found when the derivative of H_t is zero,

$$\frac{dH_t}{d\eta} = \frac{H_f}{\eta} - \frac{H_p}{\eta} = 0$$

so that $H_f = H_p$.

Viscosity is therefore optimal when the power ratio is

$$K = H_f/H_p = 1 \tag{3.14}$$

The result would be the same for aerostatic bearings. However, for an aerostatic bearing, viscosity cannot be changed so that clearance is optimized instead. In practice, it is unnecessary to take speed into account for most aerostatic bearings because air viscosity is so low. Aerostatic design is therefore often simplified.

Sometimes viscosity for a hydrostatic bearing is also constrained by other considerations. In this case too, clearance can be optimized.

Optimum Clearance h for High-Speed Bearings

Clearance is ideally chosen to be a minimum consistent with tolerances for machining, thermal expansion, and structural deflection. Selection on the basis of tolerances is not always possible, particularly if viscosity is fixed by requirements of other hydraulic equipment in the machine.

Optimum clearance is then found by varying clearance until total power from equation (3.13) is a minimum. By differentiation the optimum is found when

$$\frac{dH_t}{dh} = \frac{3H_p}{h} - \frac{H_f}{h} = 0$$

so that $H_f = 3H_p$. Optimum clearance is the value when

$$K = H_f/H_p = 3 \tag{3.15}$$

Optimum Power Ratio K *for High-Speed Bearings*

Opitz (1967) showed that with varying viscosity or clearance, total power varies by approximately 15%. Rowe and Koshal (1980) further showed for a wide range of variables that optimum power ratio, assuming zero-speed load support, always lies in the range

$$1 \leq K \leq 3 \tag{3.16}$$

As long as K is in the range indicated by condition (3.16), minimum power will not be exceeded by more than 15%.

If it were possible for the designer to vary clearance and then viscosity in turns, the process would progressively reduce both clearance and viscosity. The designer must therefore decide when to stop: decide immediately a minimum of one and optimization of the other. In practice, the range recommended provides a convenient tolerance. For a given supply pressure, the procedure has the advantage of lowest power and highest bearing stiffness.

General design points to be considered are:

1. To reduce temperature rise, it is better to reduce the land width of a bearing rather than increase the clearance. The result of doubling clearance may be compared with the result of reducing land width to a quarter of its previous value. Doubling clearance increases flow eightfold while friction shear is reduced to half. Thus, temperature rise due to friction is reduced by a factor of 16. But reducing land width to one-quarter increases flow four times, while friction is reduced to one-quarter, thus achieving the same reduction in temperature rise at half the increase in flow.
2. Decreasing viscosity has two effects: viscous shear is reduced and the relative change of viscosity with temperature is also reduced. If, say, a typical SAE 10 oil is heated from 38 to 99 °C (100–210 °F), then the viscosity changes from 37 to 6 cP, whereas a typical SAE 50 oil changes from 300 to 19 cP. Thus, the ratio of initial running torque to steady running torque is reduced with low viscosity.
3. There is a practical limit to reducing land width. If land width is too small, edge effects make the prediction of flow less accurate. Edge effects are usually negligible if land width is greater than 100 times bearing film thickness—that is, $a > 100h$. Furthermore, very thin lands are vulnerable to damage.
4. The most critical parameter governing flow is bearing gap. Often, the quickest route to a satisfactory design is therefore by optimizing gap. There is a minimum gap based on manufacturing accuracy of the bearing. Minimum gap should not be less than 10 times the Ra roughness figure for the bearing and should not be less than 2–3 times the minimum zone center roundness error. Consideration should also be given to tilt of a shaft or a bearing runner, which reduces minimum clearance at the ends of a bearing.

Optimum Pressure Ratio for High-Speed Bearings

As for low-speed bearings, gauge pressure ratio is chosen for maximum load range. The approximate pressure ratio for maximum load range in most cases is

$$\beta = K_{go} = 0.5$$

Further information on pressure ratio is given in later chapters for particular applications.

■ *Example 3.1 Calculation of Clearance for Minimum Power*

A hydrostatic bearing is to operate with minimum power. The following has already been decided by considerations such as the applied load. Oil viscosity is constrained by other hydraulic equipment.

Pressure ratio $\beta = 0.5$
Supply pressure $P_s = 3.45$ MN/m^2 (500 lbf/in^2)
Flow factor $\bar{B} = 0.5$
Oil viscosity $\eta = 34.5$ cP (5×10^{-6} reyns)
Land area $A_l = 390$ mm^2 (0.6 in^2)
Recess area $A_r = 260$ mm^2 (0.4 in^2)
Recess depth $h_r = 0.5$ mm (0.02 in)
Sliding speed $U = 2.5$ m/s (500 ft/min)

Solution
It is assumed in the calculation of friction area (see Section 3.2) that $A_r/h_r = A_r/4h$. From equations (3.1), (3.3) and (3.15) with $K = 3$,

$$h^4 = \frac{\eta^2 A_f U^2}{3 P_s^2 \beta \bar{B}} = \frac{0.0345^2 \times (390 + 260/4) \times 10^{-18} \times 2.5^2}{3 \times 3.45^2 \times 0.5 \times 0.5}$$

$$h = 0.025 \text{ mm}(0.001 \text{ in})$$

■

3.9 Speed Parameter S_h and Optimum Value S_{ho}

O'Donoghue et al. (1969) showed that the condition for minimum power from equations (3.1), (3.3), (3.4), and (3.16) can be arranged to achieve a speed parameter S_h convenient for optimum design.

The use of the design parameter S_{ho} for optimum design is *optional.* Optimal design can also be achieved simply by ensuring friction power equals pumping power or in a required proportion. Therefore, use of S_{ho} is not strictly required. However, S_{ho} is useful in adjusting combinations of design parameters, since a value of S_{ho} can be read from a chart and this value used to select values of viscosity and clearance for values of supply pressure and land-width ratio. Charts of S_{ho} values for common shapes are presented below.

The design variables are arranged into two groups.

Duty Variables

Duty variables relate to the bearing duty requirements of speed, size, supply pressure, viscosity, and clearance. The duty variables are decided initially and adjusted later as necessary.

Derived Variables

Derived variables arise from optimum performance and bearing shape considerations. The derived variables include land-width ratio, pressure ratio, power ratio, and the friction area factor.

Application of the Minimum Power Criterion

Starting from the criterion $H_f = KH_p$, where $1 < K < 3$ and substituting for H_p and H_f from equations (3.1), (3.3), and (3.4),

$$\frac{KP_s^2 \beta \bar{B} h^3}{\eta} = \frac{\eta A_f U^2}{h} \tag{3.17}$$

Plane Pads

Rearranging duty variables onto the LHS for plane pads leads to

$$S_h = \frac{\eta L U}{P_s h^2} = \sqrt{\frac{K \beta \bar{B}}{\bar{A}_f}} \qquad \text{Plane pads} \tag{3.18}$$

where $\bar{A}_f = A_f / L^2$.

The RHS of equation (3.18) includes the *derived parameters* calculated for different bearing shapes. For a given bearing shape, optimum S_{ho} has a unique numerical value corresponding to $K = 1$. Values of S_{ho} are presented in data charts such as Figure 3.1 for plane pads and Figure 3.2 for journal bearings.

Duty variables P_s, η, U, h_o, and L are adjusted until the value of the group matches S_{ho}. Varying from S_{ho} has implications for temperature rise and flow. Values of S_h higher than S_{ho} increase temperature rise. Lower values of S_h lead to increased flow. Where different values of power

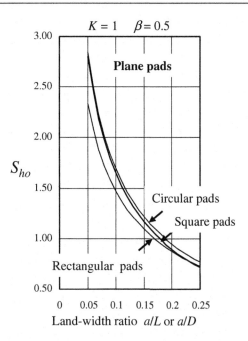

Figure 3.1: Optimum Values S_{ho} and Variations with Land-Width Ratio for Circular Pads, Square Pads ($B/L = 1$), and Rectangular Pads ($B/L = 10$).

ratio K or pressure ratio β are required, the appropriate values of S_h are found by multiplying S_{ho} by the changed values $\sqrt{K}$ or $\sqrt{\beta}$.

Values of computed shape factors for flat pads are given in Chapter 4. As an example, for recessed rectangular plane pads $S_{ho} = \sqrt{0.5\overline{B}/\overline{A}_f}$, where

$$\overline{A}_f = \frac{A_f}{L^2} = \frac{B}{L}\left[1 - \frac{3}{4}\left(\frac{B}{L} - \frac{2a}{L}\right)\left(1 - \frac{2a}{L}\right)\right]$$

Journals

In the case of hydrostatic journal bearings, the speed parameter is

$$S_h = \frac{\eta N}{P_s}\left(\frac{D}{2h_o}\right)^2 = \frac{1}{4\pi}\sqrt{\frac{K\beta\overline{B}}{\overline{A}_f}} \qquad (3.19a)$$

And for aerostatic journal bearings:

$$S_h = \frac{\eta N}{P_s}\left(\frac{D}{2h_o}\right)^2 = \frac{1}{4\pi}\sqrt{\frac{K\cdot K_{go}\overline{B}}{2\overline{A}_f}\cdot\frac{p_{ro}+p_a}{P_s}\cdot\log_e\left(\frac{p_s}{p_a}\right)} \qquad (3.19b)$$

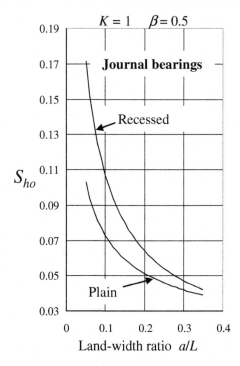

Figure 3.2: Optimized Values S_{ho} and Variations with Land-Width Ratio for Recessed and Nonrecessed Journal Bearings.

For recessed bearings:

$$\overline{A}_f = A_f/D^2 = \pi \frac{L}{D}\left[1 - \frac{3}{4}\left(1 - \frac{2a}{L}\right)\left(1 - \frac{nb}{\pi D}\right)\right]$$

and $\overline{B} = \pi D/6a$.

For nonrecessed bearings, $\overline{A}_f = \pi L/D$ and $\overline{B} = \pi D/6a$.

For maximum load range, $\beta = 0.5$. Values of the optimum speed parameter S_{ho} calculated for journal bearings are given in Figure 3.2.

Duty variables are on the LHS of equation (3.19) and *derived variables* on the RHS. For different K or β values, S_h is found by multiplying S_{ho} by the new $\sqrt{K}$ or $\sqrt{\beta}$.

S_{ho} is a unique value for a particular land-width ratio. S_{ho} ensures total power is close to minimum and is found from

$$S_{ho} = \frac{1}{4\pi}\sqrt{\frac{0.5\overline{B}}{\overline{A}_f}} \qquad \text{Hydrostatic journals} \qquad (3.20)$$

For aerostatic journal bearings, operating with $p_s/p_a = 5$, $S_{ho} = 1/4\pi(\sqrt{0.4\overline{B}/\overline{A}_f})$. In this case, the value of S_{ho} for aerostatic bearings is reduced by 10%.

■ *Example 3.2 S_{ho} for Recessed Journal Bearing*

Values specified for a recessed hydrostatic journal bearing (see Chapter 9) are as follows:

Length/diameter ratio L/D	1.0
Pressure ratio β	0.5
Land-width ratio a/L	0.25
Inter-recess land-width ratio $nb/\pi D$	0.25
Power ratio K	1.0

Calculate S_{ho} from the derived variables.

Solution

$$\overline{B} = \frac{\pi D}{6a} = \frac{\pi}{6(a/L)(L/D)} = \frac{\pi}{6 \times 0.25 \times 1.0} = 2.094$$

$$\overline{A}_f = \pi \frac{L}{D}\left[1 - \frac{3}{4}\left(1 - \frac{a}{L}\right)\left(1 - \frac{nb}{\pi D}\right)\right]$$

$$= \pi \times 1.0 \times [1 - 0.75 \times (1 - 0.25)(1 - 0.25)] = 1.816$$

$$S_{ho} = \frac{1}{4\pi}\sqrt{\frac{0.5 \times 2.094}{1.816}} = 0.06 \text{ as given by Figure 3.2}$$

■

Use of the Optimum Speed Parameter S_{ho}

In most cases it should be possible to employ a value of S_h indicated for a land-width ratio within the range $a/L = 0.1-0.25$, where 0.25 is preferred for low-speed bearings. In some cases, the value of S_h based on a minimum value of film thickness, supply pressure and viscosity, may be considerably larger than S_{ho}. Economic or other considerations may make it impracticable to reduce the viscosity in order to optimize the design to best advantage. Before increasing the film thickness, the designer should consider reducing the land-width ratio. Thus, by using a thin land-width ratio 0.1 rather than 0.25 for a rectangular pad, S_{ho} is increased from 0.72 to 1.47.

As a general rule, it is suggested that the minimum land width, a, should not be less than 50 times the film thickness h_o. If, at the minimum land width, the value of S_{ho} still falls outside an acceptable range, then h_o should be increased.

■ *Example 3.3 Optimization of a Rectangular Pad*

Consider a rectangular pad bearing, $B/L = 2$, supporting a load $W = 1.25$ kN (280 lbf). Initial values of viscosity, supply pressure, and oil film thickness are $\eta = 54.5$ cP (8 × 10^{-6} reyns), $P_s = 2.0$ MN/m^2 (290 lbf/in^2), and $h_o = 0.013$ mm (0.0005 in) respectively. Pressure ratio is $\beta = 0.5$. Velocity is $U = 0.76$ m/s (30 in/s).

Solution

Tabulated design variations are shown in Table 3.1 and demonstrate the relative merits of different optimization approaches.

If viscosity can be reduced, design 2 requires the least power but increasing power ratio to $K = 3$ as in design 3 only increases the power by 15.6% and allows higher viscosity oil to be used.

If viscosity is maintained at 54.5 cP, there are advantages of design 5 with reduced land width compared with design 4 with increased bearing film thickness. Reducing land width rather than increasing film thickness reduces total power and increases bearing stiffness.

In this example, power may seem very low for each of the design variations and unlikely to worry the designer. However, in large high-speed bearings, megawatts may be required. The design process follows the same principles irrespective of size.

■

Temperature Rise and the Minimum Power Condition

Zero-speed Bearings

The maximum temperature rise of the lubricant for a single pass through a zero-speed bearing depends only on the supply pressure, as shown by equation (3.10). Any change that reduces flow will reduce the power to be dissipated in the cooling system. There is therefore no conflict between the requirements for low temperature rise and the conditions for minimum power.

Table 3.1: Design comparisons for Example 3.3

Design	h (cP)	h (mm)	a/L	λ (MN/m)	H_p (W)	H_f (W)	H_t (W)	K
1	54.5	0.013	0.25	180	0.116	4.26	4.38	37
2	9	0.013	0.25	180	0.703	0.704	1.41	1
3	15.6	0.013	0.25	180	0.406	1.221	1.63	3
4	54.5	0.032	0.25	73	1.732	1.732	3.46	1
5	54.5	0.013	0.06	180	0.669	2.25	2.91	3.37

High-speed Bearings

The maximum temperature rise of the lubricant for a single pass through an optimized high-speed bearing is dependent only on supply pressure as shown by equation (3.9). Any change that reduces flow from the minimum power condition will increase temperature rise since H_f is increased and hence K is also increased. Minimum temperature rise corresponds to maximum flow and is achieved at the expense of increased power to be dissipated. In practice, it will be unusual for a designer to deviate from the range $1 \leq K \leq 3$ except for plain non-recessed hybrid bearings.

3.10 Optimization of Plain Nonrecessed Hybrid Bearings

The optimization of plain hybrid hydrostatic bearings of hole-entry or slot-entry types shown in Figure 1.4 follows different lines. In plain hybrid bearing design, the combined hydrostatic and hydrodynamic load support is taken into account, where the latter may be many times the former. Hydrodynamic pressures increase with power ratio and reducing land-width ratio. Increasing power ratio allows heavier bearing loads with reduced supply pressures. The designer may then use a smaller bearing that requires lower power expenditure. The direct effect of increasing K and reducing land-width ratio is to increase total power, as explained in Section 3.7. The hybrid hydrostatic bearing has therefore been optimized by a computational procedure that compares ratios of load/power. It is found that load/power may be increased by reducing land-width ratio to 0.1 and increasing power ratio within the range

$$3 \leq K \leq 9 \tag{3.21}$$

Power ratios much higher than $K = 9$ can be employed, although this introduces dangers of excessive temperature rise, hot spots, and cavitation effects that undermine bearing load support. The design of hybrid hydrostatic and aerostatic bearings is discussed further in Chapter 10. However, it is worth noting that at very high power ratios, journal bearings with central admission are less prone to hot spots than double-entry bearings.

Plain hybrid bearings operate with a higher temperature rise, according to equation (3.9).

References

O'Donoghue, J. P., Rowe, W. B., & Hooke, C. J. (1969). Design of hydrostatic bearings using an operating parameter. *Wear, 14*, 355–362.

Opitz, H. (1967). Pressure pad bearings, Conf. Lubrication and Wear, Fundamentals and Application to Design. *Proceedings of the Institution of Mechanical Engineers, London, 182*(Part 3A), 100–115.

Rowe, W. B., & Koshal, D. (1980). A new basis for the optimization of hybrid journal bearings. *Wear, 64*, 115–131.

Shinkle, J. N., & Hornung, K. G. (1965). Frictional characteristics of liquid hydrostatic journal bearings. *Journal of Basic Engineering, Transactions of the American Society of Mechanical Engineers, 87*(2; March), 163–169.

Tawfik, M., Stout, K. J., & Rowe, W. B. (1981). Characteristics of slot entry hybrid gas bearings. *Proceedings of the 8th International Gas Bearing Symposium*, Leicester Polytechnic, held in conjunction with BHRA Fluid Engineering, Cranfield.

CHAPTER 4

Pads: Area and Flow Shape Factors ($\overline{A}$ and $\overline{B}$)

Summary of Key Design Formulae

$$W = P_r A \cdot \overline{A}$$ 　　　　Load support

$$q = \frac{P_r h^3}{\eta} \cdot \overline{B}$$ 　　　　Hydrostatic flow

$$q = \frac{(p_r^2 - p_a^2) h^3}{2 \eta p_a} \cdot \overline{B}$$ 　　　　Aerostatic flow

4.1 Pad Shapes and Shape Factors

Plane thrust pads are designed to support loads applied in a single plane. This chapter provides area shape factors $\overline{A}$ and flow shape factors $\overline{B}$ for flat and curved pads.

Load support depends on the effective area A_e of a pad where $A_e = A \cdot \overline{A}$ and the total projected area of the pad is A.

Flow depends on the flow shape factor $\overline{B}$. Pad data in most cases are computed. Approximations are also given for thin-land bearings.

Aerostatic and hydrostatic pads mostly use the same shape data. Aerostatic pads and hydrostatic pads support similar loads, although aerostatic load support may be larger or smaller. For example, aerostatic load is higher by 36% for a circular pad with central admission and $R_2/R_1 = 10$. However, load support is reduced using virtual recesses, so the two effects tend to balance.

4.2 Virtual Recesses and Dispersion Losses

A *virtual recess* reduces recess volume compared with a real recess of the same size. This can be an advantage for aerostatic bearings and for some hybrid bearings. Replacing a small real

Hydrostatic, Aerostatic and Hybrid Bearing Design.
DOI: 10.1016/B978-0-12-396994-1.00004-8

recess by a large virtual recess supports more hydrostatic or aerostatic load. This is particularly useful for an aerostatic pad, improving dynamic stiffness compared with a large real recess and reducing the risk of pneumatic hammer. Stability is discussed further in Chapter 5.

A *virtual recess* is a bearing area surrounded by a ring of source areas. Each source area is fed by a restrictor, as variously illustrated in Figures 4.1–4.4 and 4.13. Pressure is approximately constant within a virtual recess if source areas provide sufficient surround. A *source area* consists of a shallow recess or shallow groove. The depth of a source area needs only to be two or three times the normal pad gap to spread source pressure almost uniformly, for example, 50–100 μm deep. A source area pattern can be produced by any convenient method, including etching, erosion, ultrasonic engraving, plating, etc.

A *surround factor SF* describes the extent to which entry sources surround a virtual recess. For example, surround factor $SF = nd_r/C_{vr}$ for circular sources or $SF = nl_g/C_{vr}$ for grooved sources, where n is the number of sources, d_r is the diameter of circular sources, and C_{vr} is the length of the periphery around the virtual recess. For grooved sources l_g is the groove length. A value $SF \approx 70\%$ usually avoids excessively reduced load support and reduced flow, except when land width, a, is much smaller than source separation distance b.

Dispersion losses reduce load support and flow from a virtual recess due to reductions in pressure around the entry sources. Dispersion is described further in the next chapter and is illustrated in Figures 4.1 and 5.21. Supply pressure or bearing area may therefore need to be increased to compensate for reduced load support.

A *dispersion loss factor QF* for reduced load and flow is roughly given by $QF \approx 0.5 + SF/2$. For example, $QF = 0.85$ with $SF = 0.7$ suggests a loss of roughly 15%. Dispersion losses are greatest with small values of SF and small values of land width a. In an extreme case, if $a < b$ then QF reduces almost to the value SF. Slot-entry bearings avoid dispersion losses.

In designing aerostatic bearings, it is suggested that maximum assumed load to be carried is increased by approximately 15% where $SF \approx 70\%$ and land width is at least equal to

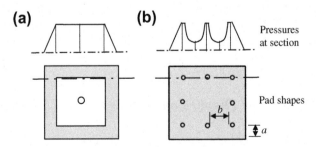

Figure 4.1: Recessed Pad (a) and a Plain Pad with a Virtual Recess (b), Showing Pressure Losses Due to Small Hole Sources.

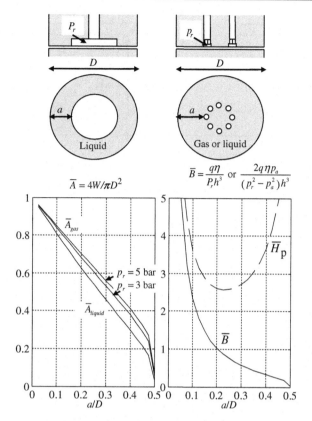

Figure 4.2: Circular Pad Data.
Suggested hydrostatic design value: $a/D = 0.25$.

source separation distance. This has the effect of increasing bearing size and compensates for loss of load support. The designer can either increase bearing size or increase supply pressure.

Surround factor *SF* can be increased by providing shallow grooves, instead of small circular recesses, into which one or more entry holes are drilled into each source area. An example of entry grooves is shown in Figure 4.3 for a square pad. Grooves have an advantage of distributing source pressure in the desired direction while reducing source volume compared with circular sources. Aerostatic source volume is discussed further in Chapter 5. For hydrostatic pads, there is no need to minimize source volume.

Virtual recesses can be applied for all hydrostatic or aerostatic pads. Land area contained within a virtual recess marginally increases friction compared with an actual recess but improves squeeze film damping and stability. Further detail on design of entry recesses for aerostatic bearings is provided in Chapter 5 when considering the effect of a recess on the restrictor value.

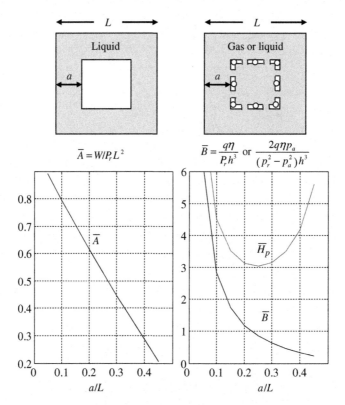

$$\bar{A} = W/P_r L^2$$

$$\bar{B} = \frac{q\eta}{P_r h^3} \text{ or } \frac{2q\eta p_a}{(p_r^2 - p_a^2)h^3}$$

Figure 4.3: Square Pad Data.
Suggested design value: $a/L = 0.25$.

■ *Example 4.1 Hydrostatic Load and Flow*

A hydrostatic pad has area $A = 0.008$ m^2 (12.4 in^2), area shape factor $\bar{A} = 0.6$ and flow shape factor $\bar{B} = 1.2$. Gauge recess pressure is $P_r = 4$ MN/m^2 (580 lbf/in^2), film thickness is $h = 25$ μm (0.001 in), and oil viscosity is $\eta = 80$ cP. Determine pad load support W and oil flow q.

Solution

$$W = P_r A \bar{A} = 4 \times 10^6 \times 0.008 \times 0.6 = 19,200 \text{ N } (4316 \text{ lbf})$$

$$q = \frac{P_r h^3}{\eta}\bar{B} = \frac{4 \times 10^6 \times 25^3 \times 10^{-18}}{0.08} \times 1.2 = 0.9375 \times 10^{-6} \text{ m}^3/\text{s } (0.056 \text{ l/min})$$

■

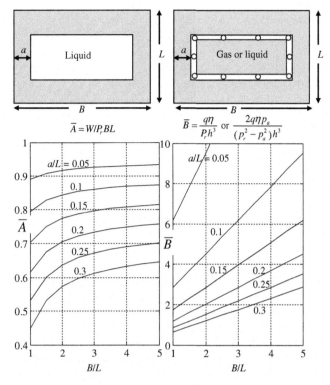

Figure 4.4: Rectangular Pad Data.
Suggested design values: $a/L = 0.25$, $1 < B/L < 2$.

■ *Example 4.2 Aerostatic Load and Flow*

An aerostatic pad has area $A = 0.008$ m² (12.4 in²), area shape factor $\bar{A} = 0.6$ and flow shape factor $\bar{B} = 1.2$. Gauge recess pressure is $P_r = p_r - p_a = 0.2$ MN/m² (29 lbf/in²). Absolute ambient pressure $p_a = 0.1$ MN/m² so that absolute recess pressure $p_r = 0.3$ MN/m². Film thickness is $h = 25$ μm (0.001 in) and oil viscosity is $\eta = 18.3 \times 10^{-6}$ N s/m² (0.002654×10^{-6} lbf s/in²). Determine load supported W and free air flow q.

Solution
Reduction in load support of a circular aerostatic pad due to partial surround of a virtual recess is offset by additional load support due to compressibility (see Figure 2.9). Load support may therefore be estimated in this case in the same way as for the hydrostatic pad. For other pad shapes, reduced load capacity may require increased supply pressure (see above).

$$W = P_r A \overline{A} = 0.2 \times 10^6 \times 0.007854 \times 0.6 = 942 \text{ N (212 lbf)}$$

$$q = \frac{(p_r^2 - p_a^2)h^3}{2\eta p_a}\overline{B} = \frac{(0.3^2 - 0.1^2) \times 10^{12} \times 25^3 \times 10^{-18}}{2 \times 18.3 \times 10^{-6} \times 0.1 \times 10^6} \times 1.2$$

$$= 410 \times 10^{-6} \text{ m}^3/\text{s} \ (\equiv 24.6 \text{ l/min})$$

∎

4.3 Circular Pads

Theory is given for a circular pad in detail to explain the approach and shape optimization. Equation (2.29) for load is

$$W = \frac{\pi P_r}{2} \frac{R_2^2 - R_1^2}{\log_e(R_2/R_1)} \tag{4.1}$$

The area shape factor is therefore given as

$$\overline{A} = \frac{W}{P_r A} = \frac{1 - R_1^2/R_2^2}{2\log_e(R_2/R_1)} \tag{4.2}$$

Equation (2.23) for hydrostatic flow is

$$q = \frac{\pi P_r h^3}{6\eta \log_e(R_2/R_1)} \tag{4.3}$$

The flow shape factor is given as

$$\overline{B} = \frac{\pi}{6\log_e(R_2/R_1)} \tag{4.4}$$

These expressions correspond to the general forms given in Sections 2.5 and 2.9:

$$W = P_r A \overline{A} \qquad\qquad \text{Load} \tag{4.5}$$

$$q = \frac{P_r h^3}{\eta}\overline{B} \qquad\qquad \text{Hydrostatic flow}$$

$$q = \frac{(p_r^2 - p_a^2)h^3}{2\eta p_a}\overline{B} \qquad \text{Aerostatic flow} \tag{4.6}$$

Effective area A_e multiplied by recess pressure P_r yields bearing load support:

$$W = P_r A_e \tag{4.7}$$

where $A_e = A\overline{A}$.

Maximum load for minimum power, as explained in Section 3.6, is the shape that minimizes

$$\overline{H}_p = \frac{\overline{B}}{\overline{A}^2} \tag{4.8}$$

Values of $\overline{H}_p$ for a circular pad are shown in Figure 4.2. The optimum is when $a/D \approx 0.24$. Small variations near an optimum make little difference, so that $a/D = 0.25$ is appropriate.

Equations (4.2) and (4.4) give $\overline{A}$ and $\overline{B}$ for a circular pad. Alternatively, $\overline{A}$ and $\overline{B}$ may be read from Figure 4.2. Values of $\overline{B}$ apply both for hydrostatic and aerostatic bearings. Hydrostatic values of $\overline{A}$ also apply approximately for aerostatic bearings. The maximum error occurs for thick-land situations. For $a/D = 0.45$, effective area is underestimated by 27% for aerostatic bearings at an absolute recess pressure of 3 bar. In other words, the error is on the safe side when using the hydrostatic value.

Flow and load support will be reduced if a virtual recess is employed rather than an actual recess. Assuming 70% surround of a virtual recess, load and flow may be reduced by approximately 15% or even more if land width $a < b$, as explained in Section 4.2.

■ Example 4.3 Circular Pad

From Figure 4.2, for $a/L = 0.25$: $\overline{A} = 0.54$ and $\overline{B} = 0.75$.

■

4.4 Square Pad Data

The geometry of the square pad and its design data are given in Figure 4.3. The length of the side is L and the land width is a. The bearing area is $A = L^2$. As for most pad shapes, the optimum land width is seen to be close to $a/L = 0.25$.

■ Example 4.4 Square Pad

From Figure 4.3, for $a/L = 0.25$: $\overline{A} = 0.53$ and $\overline{B} = 0.85$.

These values may be employed both for hydrostatic and for aerostatic pads.

For an aerostatic version, a virtual recess may be achieved employing a shallow groove linking the entry holes. Alternatively, the holes may be extended by shallow slots sufficient to ensure partial surround of a virtual recess. The effects of surround fraction are discussed in Section 4.2.

■

4.5 Rectangular Pad Data

The geometry of the rectangular pad and its design data are given in Figure 4.4. The pad area is given by $A = BL$. It is recommended that the land widths should be made equal in both directions. A suitable value of land-width ratio is given by $a/L = 0.25$.

■ Example 4.5 Rectangular Pad

From Figure 4.3, for $B/L = 2$ and $a/L = 0.25$: $\overline{A} \approx 0.63$ and $\overline{B} \approx 1.5$.

A virtual recess should be considered for an aerostatic pad to minimize air volume. An alternative is to employ a small shallow-entry recess. Reference should be made to the discussion in Section 4.2 on virtual recesses and in Chapter 5 on stability.

■

4.6 Annular Recess Circular Pad Data

Design data for an annular recess circular pad are given in Figure 4.5. This bearing is useful for taking end thrust on a collar where there is a through shaft. Pad area is given by $A = \pi(R_2^2 - R_1^2)$. The thin-land assumption gives good accuracy within 0.5% accuracy for $\overline{A}$ and makes a chart unnecessary:

$$\overline{A} = 1 - \frac{a}{R_2 - R_1} \tag{4.9}$$

A reasonable land-width ratio for a hydrostatic pad is given by $a/(R_2 - R_1) = 0.25$. Land-width ratio may reasonably be increased towards the extreme value $a/(R_2 - R_1) = 0.5$, particularly for aerostatic pads.

For an aerostatic version, reference should be made to the discussion in Section 4.2. Land-width ratio should be increased as close to 0.5 as reasonable. The groove linking the entry recesses could possibly be omitted but it is necessary to ensure an adequate surround fraction. Admission from each entry hole into an individual shallow and narrow slot is preferable to an individual circular entry recess for each hole since the surround fraction is increased and recess volume reduced.

■ Example 4.6 Annular Recess Circular Pad

From equation (4.9) and Figure 4.5:

For $R_2/R_1 = 3$ and $a/(R_2 - R_1) = 0.25$: $\overline{A} \approx 0.75$ and $\overline{B} \approx 4.2$
For $R_2/R_1 = 3$ and $a/(R_2 - R_1) = 0.5$: $\overline{A} \approx 0.5$ and $\overline{B} \approx 2.05$.

■

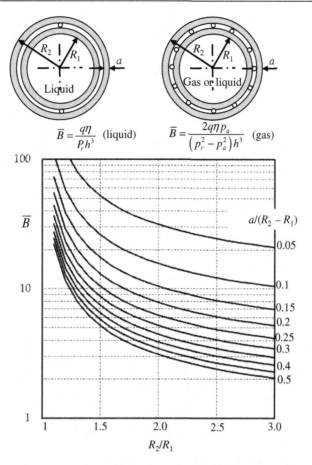

$$\overline{B} = \frac{q\eta}{P_r h^3} \quad \text{(liquid)} \qquad \overline{B} = \frac{2q\eta p_a}{\left(p_r^2 - p_a^2\right)h^3} \quad \text{(gas)}$$

Figure 4.5: Annular Recess Circular Pad: Flow Data.
Suggested design value: $a/(R_2 - R_1) = 0.25$ for liquid or 0.5 for gas.

4.7 Conical Pad Data

Conical Pad with Central Admission

A conical pad with central admission is shown in Figure 4.6. The area shape factor $\overline{A}$ for a conical pad is identical to a circular flat pad of the same projected area when viewed axially. Values of $\overline{A}$ are given in Figure 4.2, where bearing projected area is given by $A = \pi R_2^2$.

Flow shape factor $\overline{B}$ is reduced compared with a flat circular pad because of the increased length of the land due to the cone angle. The flow shape factor $\overline{B}$ from Figure 4.2 must therefore be multiplied by the sine of the half-cone angle α. Clearance h is film thickness measured perpendicularly between the two bearing surfaces. Bearing area A is defined as the projected area of the bearing viewed axially.

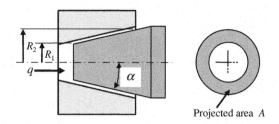

Figure 4.6: Conical Pad with Central Lubricant Admission.

For aerostatic pads, it is necessary to reduce recess volume by one of the methods discussed in Section 4.2. Design of aerostatic conical bearings is discussed more fully in Chapter 12.

■ *Example 4.7a Conical Pad with Central Admission*

From equation (4.9) and Figure 4.2:

For $R_2/R_1 = 2$ and $\alpha = 30°$: $\overline{A} = 0.54$ and $\overline{B} = 0.75 \times \sin 30° = 0.375$.

■

Conical Pad with an Annular Recess

The geometry of an annular recess conical pad is shown in Figure 4.7. The area shape factor $\overline{A}$ is the same as a flat annular bearing of the same projected area when viewed axially. Bearing projected area is given by $A = \pi(R_2^2 - R_1^2)$. For the annular recess conical pad shown in Figure 4.7, $\overline{A}$ may be obtained from equation (4.9).

Flow shape factor $\overline{B}$ is smaller than for a flat annular recess pad due to the increased length of the lands. A value of $\overline{B}$ read from Figure 4.5 must therefore be multiplied by $\sin \alpha$. Clearance h is the film thickness measured perpendicular to the bearing surfaces. Bearing area A is defined as the projected area viewed in an axial direction.

For an aerostatic annular recess pad with single-row entry, it is necessary to reduce recess volume to a minimum using a plain bearing configuration or a narrow and very shallow recess as in Figures 4.5 and 4.14.

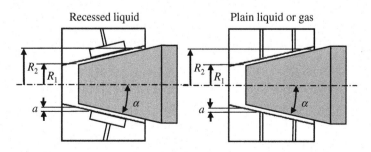

Figure 4.7: Conical Pad with Annular Entry.

■ *Example 4.7b Conical Pad with an Annular Recess*

From equation (4.9) and Figure 4.5:

For $R_2/R_1 = 2$, $\alpha = 30°$, and $a/(R_2 - R_1) = 0.2$: $\overline{A} = 0.8$ and $\overline{B} = 7.8 \times \sin 30° = 3.9$.   ■

4.8 Spherical Pad Data

A spherical pad with central admission and design data are shown in Figure 4.8. Bearing projected area is given by $A = \pi R^2$.

For an aerostatic pad it is advisable to reduce recess volume to a minimum using a virtual recess consisting of a ring of sufficient entry holes to obtain adequate surround, as discussed in Section 4.2. Stability is discussed in Chapter 5.

■ *Example 4.8 Spherical Pad with Central Admission*

From Figure 4.8, for $\alpha = 45°$, $\gamma = 90°$, $\alpha/\gamma = 0.5$: $\overline{A} = 0.8$ and $\overline{B} = 0.6$.   ■

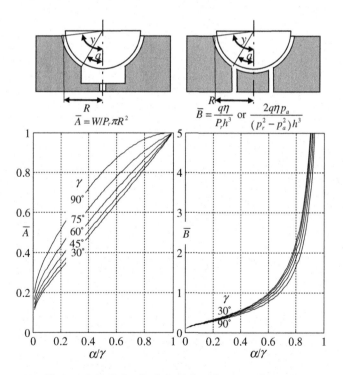

Figure 4.8: Spherical Pad Bearing Load Data.
Suggested design value: $\alpha/\gamma = 0.5$.

4.9 Multi-Recess Circular Pad Data

The geometry of a multi-recess circular pad bearing and its design data are given in Figure 4.9. Each recess has its own supply and restrictor so that multi-recess bearings can provide a small self-aligning torque when the film is nonparallel. Flow through each restrictor for a four-recess pad is one-quarter the flow for a whole pad.

The thin-land assumption works reasonably well for this configuration but slightly overestimates load and flow. Flow through each restrictor is one-quarter the flow for a whole pad.

For an aerostatic pad it is advisable to reduce recess volume to a minimum using virtual recesses surrounded by entry holes to obtain adequate surround, as discussed in Section 4.2. Stability is discussed in Chapter 5.

Bearing area is given by $A = \pi R^2$.

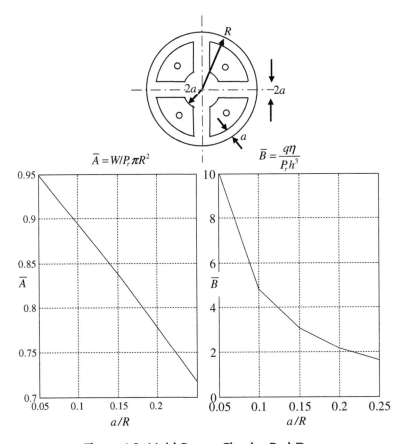

Figure 4.9: Multi-Recess Circular Pad Data.
Suggested design value: $0.15 < a/R < 0.25$.

■ Example 4.9 Multi-Recess Circular Pad

From Figure 4.9, for $a/R = 0.2$: $\overline{A} = 0.78$ and $\overline{B} = 2.2$.

■

4.10 Multi-Recess Rectangular Pad Data

The geometry of a multi-recess rectangular pad bearing and its design data are given in Figure 4.10. Each recess is supplied through its own restrictor so that the bearing has a self-aligning tendency when the film is nonparallel. Flow through each restrictor is one-quarter the total flow given by $\overline{B}$ for a four-recess pad. Bearing area is given by $A = BL$.

For an aerostatic pad it is advisable to reduce recess volume to a minimum using virtual recesses consisting of sufficient entry holes to obtain adequate surround, as discussed in Section 4.2. Stability is discussed in Chapter 5.

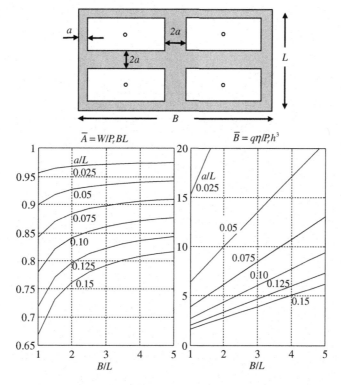

Figure 4.10: Multi-Recess Rectangular Pad.
Suggested range: $1 < B/L < 2$; $a/L = 0.125$.

■ *Example 4.10 Multi-Recess Rectangular Pad*

From Figure 4.10, for $B/L = 2$ and $a/L = 0.15$: $\overline{A} \approx 0.76$ and $\overline{B} \approx 2.7$.

■

4.11 Data for Rectangular Pad with Radiussed Recess Corners

The geometry of a rectangular pad with radiussed corner recesses and corresponding design data are given in Figure 4.11. This configuration is useful when the recess is produced using an end mill. Calculation follows the same lines as Example 4.4. Data are taken from Figure 4.11 and are given both for a radius $r = 0.5a$ and for a radius $r = a$. The effect of the radius is seen to be a reduction in bearing load support and flow as indicated by $\overline{A}$ and $\overline{B}$.

For an aerostatic pad it is advisable to reduce recess volume to a minimum using a virtual recess consisting of a ring of entry holes to obtain adequate surround, as discussed in Section 4.2. Stability is discussed in Chapter 5. Bearing area is given by $A = BL$.

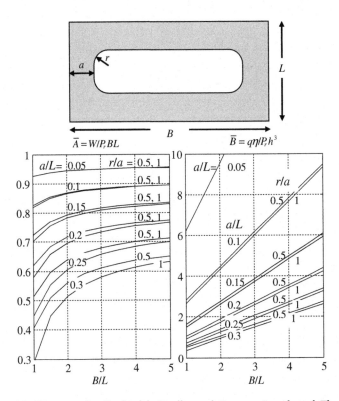

Figure 4.11: Rectangular Pad with Radiussed Recess: Load and Flow Data.

■ *Example 4.11 Rectangular Pad with Radiussed Recess Corners*

From Figure 4.11, for $B/L = 2$, $r/a = 0.5$, $a/L = 0.25$: $\overline{A} = 0.63$ and $\overline{B} = 1.3$.

■

4.12 Data for Any Shape with Thin Constant Land Width

Effective area, A_e, may be obtained very easily for any thin-land bearing assuming that recess pressure is effective over the recess area and half the land area. This assumption gives good accuracy for a reasonably regular non-re-entrant shape when a/L is less than or equal to 0.1.

The flow factor may be estimated approximately on the basis that the flow takes place through a slot of width equal to the perimeter around the mid-land and that the length in the direction of flow is equal to the land width a.

Expressions for $\overline{A}$ and $\overline{B}$ are given in Figure 4.12. These expressions provide a cross-check on the order of magnitude of computed data.

Application is by the same method as for previous examples above.

■ *Example 4.12 Any Thin-Land Shape*

From Figure 4.12, for recess area/total area $= 0.8$, total area $A = 0.1$ m^2, and $a = 0.03$ m: $\overline{A} \approx 0.9$ and $\overline{B} \approx 1.85$.

■

4.13 Annular Multi-Recess Pad Data

The geometry of this bearing pad shape is given in Figures 4.13 and 4.14. The multi-recess bearing generates a self-aligning torque when the bearing surfaces are nonparallel. Each

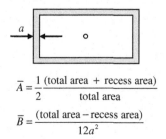

Any shape with thin constant land width

$$\overline{A} = \frac{1}{2}\frac{(\text{total area } + \text{ recess area})}{\text{total area}}$$

$$\overline{B} = \frac{(\text{total area} - \text{recess area})}{12a^2}$$

Figure 4.12: Thin-Land Bearing Data for Load and Flow.

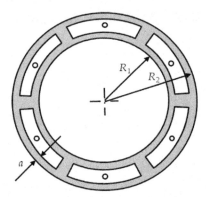

Figure 4.13: Hydrostatic Multi-Recess Annular Pad.
See Figure 4.5 for flow data. Suggested design range $0.1 < a/(R_2 - R_1) < 0.3$.

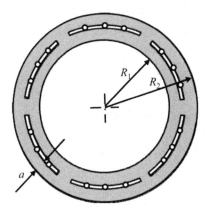

Figure 4.14: Aerostatic Multi-Recess Annular Pad.
See Figure 4.5 for flow data. Suggested design value $a/(R_2 - R_1) \approx 0.5$.

recess is supplied with its own restrictor, so that flow through one restrictor of a six-recess pad is exactly one-sixth of the total flow.

$\overline{A}$ and $\overline{B}$ for total load and flow may be taken from equation (4.10) and Figure 4.5. These values are reasonably accurate for the multi-recess case because the lands of an annular recess bearing are inherently thin. Values of $\overline{A}$ and $\overline{B}$ will be slightly reduced if the inter-recess land is disproportionately large.

Bearing area is given by $A = \pi(R_2^2 - R_1^2)$.

$$\overline{A} \approx 1 - \frac{a}{R_2 - R_1} \tag{4.10}$$

For an aerostatic version, reference should be made to the discussion in Section 4.2. The land-width ratio for the annular multi-recess pad should be increased as closely to 0.5 as physically possible (see Figure 4.14). The groove linking the entry recesses could possibly be omitted but it is necessary to ensure an adequate surround fraction. Admission from each entry hole into an individual shallow and narrow slot is preferable to an individual circular entry recess for each hole since the surround fraction is increased and recess volume reduced. In this example three restrictors are shown for each recess to ensure fast response of the recess pressures. While the thickness of the slot recess is very small, the arc covered by the slot is large and ensures that the pressure distribution does not suffer from large pressure drops due to a large space between the restrictors.

A reasonable land-width ratio for a hydrostatic pad is given by $a/(R_2 - R_1) = 0.25$. Land-width ratio for an aerostatic pad should be increased to minimize air volume (see Figure 4.14).

■ *Example 4.13 Annular Multi-Recess Pad*

From Figure 4.5, for $R_2/R_1 = 3$ and $a/(R_2 - R_1) = 0.25$: $\overline{A} \approx 0.75$ and $\overline{B} \approx 4.2$.

From Figure 4.5, for $R_2/R_1 = 3$ and $a/(R_2 - R_1) = 0.5$: $\overline{A} \approx 0.5$ and $\overline{B} \approx 2.0$.

■

CHAPTER 5

Flow Control and Restrictors

Summary of Key Design Formulae

$$\overline{X} = \frac{h}{h_o}$$ Bearing film thickness

$$\beta = K_{go} = \frac{P_{ro}}{P_s}$$ Gauge pressure ratio when $h = h_o$

$$A_e = A \cdot \overline{A}$$ Effective pad area

$$\lambda = \frac{P_s A_e}{h_o} \cdot \overline{\lambda}$$ Bearing film stiffness

5.1 Introduction

Hydrostatic and aerostatic bearings require a flow control so that film pressures increase or decrease with applied load. Each recess or entry that carries load independently must include a separate control. A simple example is a capillary restrictor that varies pressure in proportion to flow. As film thickness decreases under increased load, flow is reduced, and entry pressure increases until equilibrium is reached.

A bearing withstands a range of loads if supply pressure is much larger than recess pressure. Supply pressure higher than recess pressure allows recess pressure to either increase or decrease with applied load. Load range can be engineered by adjustment of the design pressure ratio, termed β for liquids or K_{go} for gases.

In this chapter, various flow controls are described along with effects of flow control characteristics on bearing performance. For simplicity, initial consideration is mainly focused on hydrostatic flow control. Further discussion of aerostatic flow control follows in Section 5.7, including application of virtual recesses.

Flow control particularly affects bearing film stiffness at the set design gap. Average stiffness λ can be estimated based on the load $W_{0.5}$ supported at a deflection from the design gap equal to a fraction of the film thickness. For example, for 50% deflection $\lambda \approx 2W_{0.5}/h_o$.

Hydrostatic, Aerostatic and Hybrid Bearing Design.
DOI: 10.1016/B978-0-12-396994-1.00005-X

5.2 Bearing Film Stiffness

The need for a flow restrictor may be considered for variations in applied load. Load support by a hydrostatic or aerostatic pad is $W = P_r A_e$. A bearing without a flow restrictor either lifts completely off its pad for recess pressure greater than required to support the load, or collapses for recess pressure less than required. Varying applied loads require control of flow and pressure to stabilize variations in film thickness. In other words, a bearing requires "film stiffness". Film stiffness is defined as $\lambda = -dW/dh$.

The action of a flow control is illustrated in Figure 5.1. The pressure distributions show a supply pressure P_s, a low recess pressure P_{r1} for a low applied load, and an increased recess pressure P_{r2} for a high applied load. The film thickness h between the bearing lands and the opposite bearing surface is shown greatly exaggerated.

Lubricant passes through a restrictor and enters the bearing recess at recess pressure P_r. The bearing film thickness h depends on recess pressure and flow. When bearing load W is increased, recess pressure is increased and flow through the restrictor is reduced. Film thickness is reduced accordingly. In this way, a restrictor controls the changes in flow with recess pressure. In performing this function, a restrictor provides film stiffness.

The importance of high stiffness is discussed for specific applications in Section 1.3. For many users, high stiffness is a prime reason for selecting hydrostatic or aerostatic bearings. Average

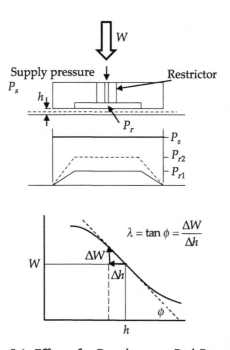

Figure 5.1: Effect of a Restrictor on Pad Pressures.

film stiffness is given by $\lambda \approx$ force/deflection. The differential form $\lambda = -dW/dh$ is more correct for fluid films because the load–deflection relationship is nonlinear, as shown in Figure 5.1.

Total deflection of a bearing assembly due to an applied force is the change in film thickness plus the change in structural deflection, $dx_t/dW = dx_s/dW + dh/dW$. Total stiffness is therefore given by $1/\lambda_t = 1/\lambda_{spring} + 1/\lambda_{film}$. A designer may wish to know the deflection x, caused by a small additional force F added to a load W. As long as F remains within a linear range on the load–deflection curve, deflection is given by $x = F/\lambda$.

5.3 Hydrostatic Circuit Design and Sealing

The simplest method of flow control is a capillary restrictor in the supply line. A capillary, as shown in Figure 5.1, produces a pressure drop proportional to flow. Circuit design as shown in Figure 5.2 also involves aspects of pumps, valves, filters, and seals.

Typically, a fixed displacement pump draws lubricant from a tank through a strainer and delivers it through a line filter for each flow restrictor at a pressure P_s determined by a pressure-relief valve. A filter in the return line from the pressure-relief valve allows debris to be constantly removed from the lubricant before and during the time the bearing is connected to the bearing. It is important to ensure that restrictors are not blocked, leading to bearing failure. If restrictors are removable, cleaning is more convenient and maintenance is easier.

Bearing oil may drain back to a tank under gravity, or alternatively it may be collected in a header tank under the bearing and pumped back to the main tank if the level in the header

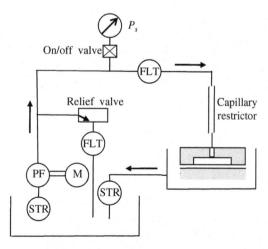

Figure 5.2: A Circuit for a Capillary-Controlled Single Plane Pad.

tank rises. This requires a switch to sense oil levels in the header tank so that the pump does not continue to operate after the header tank is emptied.

The pump required to supply a bearing depends on the supply pressure. Usually, pumps for low pressure are quieter and less costly than those for high pressure. Screw pumps when available may be quiet compared with fixed displacement pumps.

An oil return system also requires consideration of sealing. This is important to prevent ingress of damaging abrasive particles. Loss of the lubricant or contamination of other processes is also important. Basically there are five possibilities:

1. *Total loss system.* It may be permissible in some systems to discard the leakage from the bearing. This is obviously the system employed for air bearings and sometimes for water-lubricated systems. A total loss system is very attractive since it is straightforward to drain effluent to a convenient and acceptable disposal area. Seals can be a source of many problems in an otherwise ideal bearing. Sealing problems are eliminated in a total loss system.
2. *Completely enclosed system.* In some cases, it is possible to completely cover the whole bearing area with an apron or bellows, which traps the lubricant and ensures that there is neither escape of lubricant nor ingress of foreign matter.
3. *Labyrinth system.* Most of the lubricant is drained back to the tank. The labyrinth is a noncontact system that allows some loss leakage to occur through a high resistance flow path. In this way, loss leakage is reduced to a very small value. The labyrinth requires careful machining.
4. *Rubbing seals.* There are a variety of rubbing seals and care should be taken in their selection as the seals may undermine the high quality of bearing location and smoothness achieved by the bearings. If the pressure has been reduced to ambient pressure by careful design, the seal forces required are minimal and wear should be negligible over long periods.
5. *Pressure seals.* In some applications, liquid loss may be prevented by the application of low air pressure in the zone that must be sealed. This pressure will be sufficient to force the lubricant through the normal drain leading back to the tank but not so high as to cause excessive aeration of the lubricant.

Aerostatic flow control and circuit design are described further in Section 5.7.

5.4 Load and Stiffness of Capillary-Controlled Pads

Pressure drop across a hydrostatic pad capillary restrictor of length l_c and internal diameter d_c (Section 2.6) is given by $P_s - P_r = K_c \eta q$, where $K_c = 128 l_c / \pi d_c^4$ is calculated or read directly from Figure 5.3. A greater range of values is easily obtained from the chart since reducing d_c by

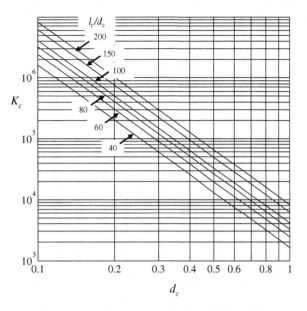

Figure 5.3: Chart for Capillary Factor K_c.

a factor of 10 increases K_c by 1000 times. The same procedure applies for aerostatic bearings. Bearing film stiffness is derived starting from equation (2.30):

$$W = \frac{P_s A_e}{1 + K_c \overline{B} h^3} \qquad \text{Liquid}$$

$$W = \left[\sqrt{\frac{p_s^2 + p_a^2 K_c \overline{B} h^3}{1 + K_c \overline{B} h^3}} - p_a \right] A_e \qquad \text{Gas} \tag{5.1}$$

Film stiffness is therefore:

$$\lambda = -\frac{dW}{dh} = P_s A_e \frac{3 K_c \overline{B} h^2}{[1 + K_c \overline{B} h^3]^2} \qquad \text{Liquid}$$

$$\lambda = -\frac{3 K_c \overline{B} h^2 (p_s^2 - p_a^2) A_e}{2 \left[\frac{p_s^2 + p_a^2 K_c \overline{B} h^3}{1 + K_c \overline{B} h^3} \right]^{1/2} [1 + K_c \overline{B} h^3]^2} \qquad \text{Gas} \tag{5.2}$$

Dimensionless Stiffness Related to the Design Condition $h = h_o$

Simplifications apply at the design condition, as shown by the following. Parameters at the design condition are denoted as h_o, P_{ro}, q_o, and β. These are values of film thickness, recess pressure, flow rate, and pressure ratio at a single design condition. Design pressure ratio β for hydrostatic bearings or K_{go} for aerostatic bearings and film thickness ratio $\overline{X}$ are defined as

$$\beta = K_{go} = \frac{P_{ro}}{P_s} \quad \text{and} \quad \overline{X} = \frac{h}{h_o}$$

Equating flows for capillary control:

$$q_o = \frac{(1-\beta)P_s}{K_c\eta} = \frac{P_s\beta\overline{B}h_o^3}{\eta} \qquad \text{Liquid}$$

it follows that

$$K_c\overline{B}h_o^3 = \frac{1-\beta}{\beta} \qquad \text{Liquid}$$

$$K_c\overline{B}h_o^3 = \frac{1-K_{go}}{K_{go}} \cdot f_k \qquad \text{Gas}$$

(5.3)

The factor f_k for an aerostatic bearing is

$$f_k = \left[\frac{1 + p_s/p_a + K_{go}(p_s/p_a - 1)}{2 + K_{go}(p_s/p_a - 1)} \right] \tag{5.4}$$

For example, if $K_{go} = 0.5$ and $p_s/p_a = 5$, then $f_k = 2$. This implies that a restrictor for gas flow must have a smaller gap or a longer flow path than for liquid flow. A typical capillary is twice as long or has 16% smaller diameter.

At the design condition, equations (5.1) reduce to similar forms for liquid and for gas:

$$W_o = P_sA_e\beta \quad \text{or} \quad W_o = P_sA_eK_{go}$$

Moving away from the design condition, load and stiffness are

$$W = \frac{P_sA_e}{1 + [(1-\beta)/\beta]\overline{X}^3} \qquad \text{Liquid}$$

$$W = \left[\sqrt{\frac{p_s^2 + p_a^2[f_k(1-K_{go})/K_{go}]\overline{X}^3}{1 + [f_k(1-K_{go})/K_{go}]\overline{X}^3}} - p_a \right] A_e \qquad \text{Gas}$$

(5.5)

$$\lambda = \frac{3P_sA_e}{h_o} \frac{\beta}{1-\beta} \frac{\overline{X}^2}{[\overline{X}^3 + \beta(1-\beta)]^2} \qquad \text{Liquid}$$

$$\lambda = \frac{3P_sA_e}{2h_o} \frac{m_k\overline{X}^2(p_s + p_a)}{\left[\sqrt{\frac{p_s^2 + p_a^2 m_k\overline{X}^3}{1 + m_k\overline{X}^3}} \right][1 + m_k\overline{X}^3]^2} \qquad \text{Gas}$$

(5.6)

where

$$m_k = f_k(1 - K_{go})/K_{go} \tag{5.7}$$

The above equations are normalized using $W = P_s A_e \overline{W}$ and $\lambda = (P_s A_e/h_o)\overline{\lambda}$. When $h = h_o$, $\overline{X} = 1$ and dimensionless stiffnesses for capillary control are given by

$$\overline{\lambda}_o = 3\beta(1 - \beta) \qquad\qquad \text{Liquid}$$

$$\overline{\lambda}_o = 3\frac{m_k}{[1 + m_k]^2}\frac{p_s + p_a}{2p_{ro}} \qquad \text{Gas} \tag{5.8}$$

For example, if $\beta = 0.5$, $\overline{\lambda}_o = 0.75$ for a hydrostatic pad. For an aerostatic pad, with $K_{go} = 0.5$ and $m_k = 2$, $p_{ro}/p_a = 3$ and $\overline{\lambda}_o = 0.667$. Dimensionless stiffness is reduced by approximately 11% for an aerostatic pad with capillary control.

$\overline{W}$ and $\overline{\lambda}$ are presented in data charts in Chapter 7 for plane pads. Relationships and data charts are presented in Chapter 9 for journal bearings. The following example calculates bearing stiffness for a circular thrust pad.

■ *Example 5.1 Load and Stiffness for a Given Pressure Ratio*

It is required to calculate bearing thrust and stiffness in the design condition for a circular hydrostatic thrust bearing of effective area $A_e = 1600$ mm^2 (2.48 in^2). The length and diameter of the capillary restrictor have been chosen so that $P_r = 0.5P_s$ when $h = h_o = 25$ μm (0.001 in). In other words, $\beta = 0.5$. A supply pressure of 3 MN/m^2 (435 lbf/in^2) is set by a relief valve. The bearing flow factor is $\overline{B} = 0.75$. The bearing is pressurized with oil of viscosity $\eta = 34.5$ cP (5×10^{-6} reyn) at 38 °C.

Solution
From equation (5.5),

$$W = \frac{3 \times 10^6 \times 1600 \times 10^{-6}}{1 + \dfrac{1 - 0.5}{0.5} \times 1^3} = 2.4 \text{ kN (540 lbf)}$$

From equation (5.6),

$$\lambda = \frac{3 \times 3 \times 10^6 \times 1600 \times 10^{-6}}{25 \times 10^{-6}} \cdot \frac{0.5}{1 - 0.5} \cdot \frac{1^2}{\left[1^3 + \dfrac{0.5}{1 - 0.5}\right]^2}$$

$$= 144 \text{ MN/m (822,000 lbf/in)}$$

Deflection can be related to stiffness using: *deflection = load/stiffness* for small load changes. For example, an added load of 100 N (22.5 lbf) placed on the supported member will cause a deflection $100/144 \times 10^{-6} = 0.7$ μm (27 mic. in).

■

5.5 Flow Control Mechanisms

Flow Characteristics

Flow control devices of various types are used in hydrostatic and aerostatic bearings. Each type yields different film stiffness. Control types include:

- Laminar restrictors: such as capillary, slot, or annular restrictors
- Turbulent restrictors: such as an orifice of any shape
- Constant-flow control: various valves or pumps
- Pressure-sensing valves
- Inherent flow control: such as shallow recess or inherent annular orifice restrictors.

The first four are listed in order of increasing stiffness. These four are external flow control types.

Inherent flow control does not require an external device but relies on pad shape and internal restriction to achieve film stiffness. Inherent flow control usually yields lower stiffness than external control devices.

A further form of inherent flow control is the Zollern patented gap compensation system. This ingenious flow control mechanism is featured in opposed-pad hydrostatic products supplied by Zollern GmbH. Each pad recess contains a gap-sensing throttle land fed with supply pressure. The flow through the throttle land is controlled by the bearing gap. The flow out from the throttle is then fed through a communicating channel to the load-carrying recess in the opposing hydrostatic pad. Therefore, flow to the pad with increased gap tends to be reduced while flow to the pad with reduced gap tends to be increased. Bearing gap control is achieved without recourse to capillary, slot, or orifice restrictors. The system is known as "bearing clearance" compensation. Flow control and concentric pressure ratio employing this system depend on appropriate throttle pad areas and load-carrying areas.

Film stiffness depends on the pressure-flow characteristics of the flow control. Pressure-flow characteristics are illustrated in Figure 5.4 for incompressible flow.

Capillaries and other laminar-flow restrictors give least stiffness whereas optimized pressure-sensing valves give greatest stiffness. A wrongly designed pressure-sensing valve may lead to negative stiffness and hence static instability. Figure 5.4 shows the optimum slope is unity.

Laminar-Flow Restrictors: Capillary, Slot, Annulus, and Cone

Examples of laminar-flow restrictors are illustrated in Figure 5.5, together with the corresponding hydrostatic flow expressions for capillary, slot, annulus, and cone restrictors. Although stiffness, with laminar-flow restrictors such as capillary, slot and annular restrictors,

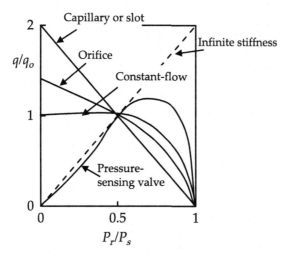

Figure 5.4: Flow and Pressure Characteristics of Various Flow Control Devices.

is less than with other forms of external restrictor, this group is often preferred since stiffness is more linear over the load range and stiffness of hydrostatic bearings is usually excellent. There are also other advantages related to reliability. Example 5.1 on page 89, shows that high stiffness can be achieved with moderate supply pressure and bearing area.

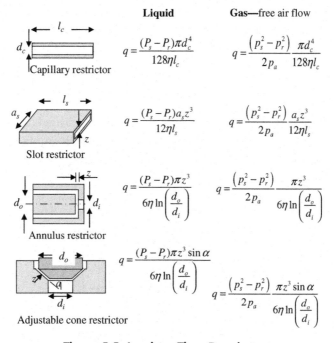

Figure 5.5: Laminar-Flow Restrictors.

A major advantage of simple laminar-flow restrictors is that they endow the bearing with the greatest tolerance to manufacturing variations on bearing clearance and to variations in operating temperature.

Capillary restrictors have three advantages. The first is manufacturing simplicity, since a capillary restrictor is made by cutting a tube of known bore to a suitable length. The second, as with other laminar-flow restrictors such as slot restrictors, is that bearing load and stiffness are independent of fluid viscosity and hence temperature rise; this can be verified by examination of equations (5.5) and (5.6). For other forms of flow control, variation of temperature causes the bearing to adopt a different film thickness for a given bearing thrust. The third advantage applies equally to hydrostatic and aerostatic bearings: stiffness remains more nearly constant across the load range for varying film thickness. It is not usually practicable to apply capillary-tube restrictors for aerostatic bearings due to the very small dimensions required. However, laminar-flow slot restrictors make an ideal alternative.

Four common methods of making restrictors are: (1) hypodermic tubing (commercially available); (2) glass capillary (commercially available); (3) hole drilling; (4) spark machining.

Methods 3 and 4 do not usually produce pure capillary action since the length-to-diameter ratio achievable is insufficient. This means that end faces must be carefully finished to control end effects if matched flow characteristics are required for a set of bearing pad restrictors. It also necessitates calibration. End effects are negligible only if the length-to-diameter ratio exceeds 100. However, often a smaller ratio will be adopted.

Equation (2.19) for flow through a capillary applies only to laminar flow. Therefore, it is necessary to check that the Reynolds number is less than 1000.

There have also been various designs of adjustable length capillaries or slots based on the screw-thread concept. One example is shown in Figure 5.6.

Slot restrictors are ideal for surrounding a pressurized region, particularly in journal bearings. Slots avoid the need for bearing recesses where slots surround a required region. Such regions are termed "virtual recesses" (see Section 4.2). Virtual recesses are invaluable for aerostatic

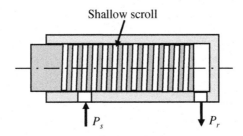

Shallow scroll

P_s P_r

Figure 5.6: Adjustable Length Capillary/Slot.

journal bearings but may also be employed for plain hydrostatic journal bearings. The width of the slots should surround at least 75−85% of the pressurized region.

Turbulent-Flow Restrictors: Orifices

Liquid flow through an orifice is related to the pressure difference across it by $q = K\sqrt{P_s - P_r}$. Expressions for liquid and gas flow are shown in Figure 5.7 and described in Chapter 2.

Unlike a laminar-flow capillary-controlled bearing, a turbulent-flow orifice-controlled bearing is temperature dependent, because there is no viscosity term in the restrictor flow equation to balance the effect of viscosity on flow through the bearing lands.

Orifices are more compact than capillaries and give fractionally greater stiffness, so that for light loadings and bearings of high stiffness the dependence on temperature may not be critical. Rippel (1963) suggested that orifices are more prone to silting, leading to changes in the orifice characteristic; blockage of slots or orifices leads to bearing failure.

Figure 5.7 shows typical orifice restrictors of length less than diameter. Ideally, orifice diameter should be less than 0.1 times the diameter of the tube supplying the orifice. The flow factors for an orifice depend on the Reynolds number, as indicated in a typical chart. Due to the very short

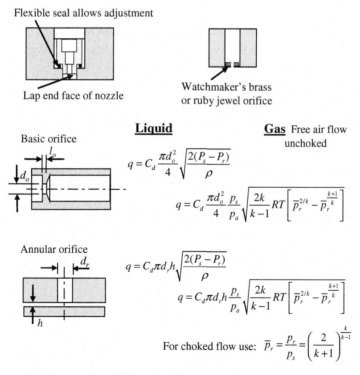

Figure 5.7: Orifice Restrictors: Basic and Annular.

length of an orifice, turbulent restriction is achieved at much lower values of *Re* than in a long capillary. An orifice nozzle is sometimes manufactured within a plug of suitable material such as a high-quality machining brass. Nozzles have been successfully manufactured where the plug body can be adjusted for position, allowing the pocket depth to be accurately set. Nozzle flow of the jets should be calibrated prior to assembly. Nozzles have also been made by pressing or gluing watchmaker's ruby or brass jewel orifices flush into the face of a drilled plug. Care must be taken to ensure that orifices are not closed up when pressed in place. Also, the orifices and the faces should be free from burrs.

Annular orifices are sometimes employed for hole-entry hybrid bearings or for shallow-recess bearings. An annular orifice gives bearing film stiffness that is two-thirds the bearing film stiffness of a simple orifice bearing. Whereas a basic orifice restriction is the area $\pi d_o^2/4$, an annular orifice restriction is the area $\pi d_r h$. An annular orifice is sometimes designed such that the flow enters directly into the clearance of a nonrecessed bearing and is not throttled within the entry hole. An annular orifice becomes effective when the annular entry area $\pi d_r h$ is less than the throat area $\pi d_r^2/4$ of the entry hole. In some cases, particularly for aerostatic bearings, the restriction may depend on both a basic orifice area and an annular orifice area, as explained in Section 5.7. A factor δ may be introduced to take account of the combined effect.

Constant-Flow Control

Constant-flow control systems are illustrated in Figure 5.8. Constant-flow control only requires a constant-flow pump or a constant-flow valve in the supply line for a single pad. However, for a large number of recesses, the cost of a number of pumps or valves can make constant-flow control an expensive option. A maximum supply pressure must be ensured at which constant flow can be maintained. The maximum supply pressure governs the maximum load that can be applied to the bearing, as for other types of flow control.

It appears at first sight that constant-flow control offers better stiffness and load-bearing capacity than the previous two flow control restrictor types considered. However, in practice, performance gains are limited by two factors:

1. There is a maximum pressure for a pump at which constant flow can be maintained. It is usually necessary to limit maximum pressure by adjustment of the relief valve.
2. There is a minimum pressure difference across a valve for satisfactory operation. This will not usually be less than 0.2 MN/m^2 (30 lbf/in^2).

Pressure-Sensing Valves

There are two main groups of pressure-sensing valves based either on spool valves or on diaphragm valves. In both systems the flow restriction is varied according to the load on the bearing. An optimum pressure-sensing valve increases flow rate in proportion to bearing load.

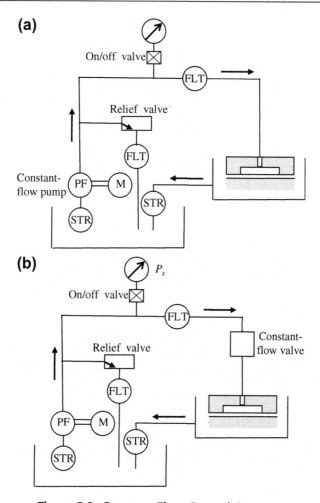

Figure 5.8: Constant-Flow Control Systems.
(a) A constant-flow pump. (b) A constant-flow valve for each recess.

This maintains film thickness constant within the limitations of system linearity and yields virtually infinite static stiffness of the bearing film. The Rowe diaphragm valve was developed for journals and other opposed-pad bearings (Rowe, 1966/1969; Rowe and O'Donoghue, 1969–70).

The application of pressure-sensing valves is usually restricted to hydrostatic applications. Although pressure-sensing valves have been employed for aerostatic bearings, the problems of ensuring stability of aerostatic bearings tend to discourage their application.

Spool Valves

Royle et al. (1962) invented a valve shown diagrammatically in Figure 5.9 in which recess pressure P_r remains equal to half valve pressure P_v. Consequently, flow increases in proportion

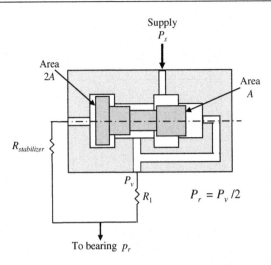

Figure 5.9: Royle Valve for Infinite Stiffness.

to recess pressure, the condition for infinite stiffness. The bearing has infinite stiffness up to the point that $P_r > P_s/2$. For greater bearing loads, the system acts as though capillary compensated. Resistance R_1 is set equal to the bearing resistance at the required film thickness $h = h_o$. Spool valves give a slower response than diaphragm valves due to the mass of the spool. There is also a danger of sticking motion of the valve, which must be avoided.

Diaphragm Valves

Diaphragm valves are inherently fast acting and simple to manufacture. The following are two examples.

Mohsin valve (Mohsin, 1962)

The Mohsin valve was designed to provide infinite stiffness of a single hydrostatic pad. One form of this valve is shown in Figure 5.10. The principle of operation is that an increase in bearing load increases the pocket pressure P_r, which in turn increases the force exerted on the diaphragm by the lubricant. Under increased pressure, the diaphragm deflects away from the supply pressure source, thus reducing the controlling restriction. With careful selection of the diaphragm stiffness, diaphragm deflection causes a proportional increase in flow and hence infinite stiffness can be obtained over a large part of the loading curve.

The coil spring, which acts against the diaphragm, does not contribute significantly to the spring rate of the diaphragm. Its function is to preload the diaphragm against the normal operating oil pressure. If the diaphragm is too thin for the application, static instability results.

A limitation of the Mohsin valve is that it cannot be readily applied to an opposed-pad bearing. Applying a Mohsin valve to each side of the opposed-pad bearing causes each valve to attempt

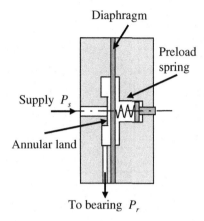

Figure 5.10: Mohsin Valve for Infinite Stiffness.

to gain control of film thickness. Either both valves become fully closed or both valves become fully open and the result is low or zero stiffness. Mohsin overcame this problem by using an adding valve. The adding valve ensures that the sum of the opposing recess pressures is approximately equal to the supply pressure.

The above limitation for opposed-pad bearings was overcome in another way by the Rowe valve.

Rowe valve (Rowe, 1966/1969)

An arrangement of the Rowe valve for opposed pads is shown in Figure 5.11 and for journal bearings in Figure 5.12. The diaphragm is clamped symmetrically and the undisturbed pressures balance on each side, avoiding the need for a preload mechanism. The construction is inherently simple and a number of pads can be controlled from one diaphragm block having the appropriate number of restrictors incorporated into the surface. In an actual valve incorporated into a grinding machine spindle, 14 bearing pads were controlled from a single valve block. This provided an economic manufacturing solution.

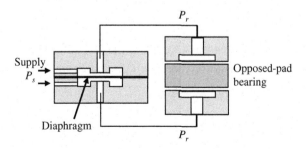

Figure 5.11: Rowe Valve for Opposed-Pad Bearings of Infinite Stiffness.

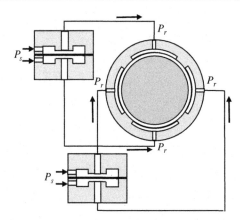

Figure 5.12: Rowe Valve for Journal Bearings.

Optimized diaphragm stiffness yields a bearing of infinite stiffness. On each side of the diaphragm, lubricant flows from an annular ring to a central supply hole, which leads to the bearing recess. A load on the bearing causes a deflection of the diaphragm, which increases flow on the loaded side and reduces flow on the unloaded side. An unloaded bearing adopts a central position, irrespective of the manufactured clearance.

Tests under laboratory-controlled conditions showed the following advantages over capillary compensation: (1) faster response; (2) larger bearing tolerances permissible; (3) greater load-bearing capacity.

Typical load–deflection characteristics of a diaphragm-controlled plane pad bearing are shown in Figure 5.13. As with all diaphragm valves, the thickness of the diaphragm must be selected for the application in order to achieve optimum stiffness. It can be seen that if the diaphragm is too thick, the bearing characteristics are identical to a capillary-controlled bearing. Adjusting the diaphragm thickness for optimum stiffness yields infinite stiffness over almost half the load range. If the diaphragm is too thin the bearing moves towards the applied load, which is defined as negative stiffness or greater than infinite stiffness. This condition is not recommended as it can lead to static instability and limit-cycle oscillations. If limit-cycle oscillations are experienced, the cure is to increase diaphragm thickness or reduce supply pressure.

Reference Bearings

It may be desirable in some applications to control the position of the main load-supporting member from a small, precisely manufactured, reference bearing.

Mohsin Reference Bearing System (Mohsin, 1962)

Figure 5.14 shows an arrangement employing a diaphragm valve and a relay valve. With careful design, the whole system is positioned with the accuracy of the reference bearing system and with infinite stiffness.

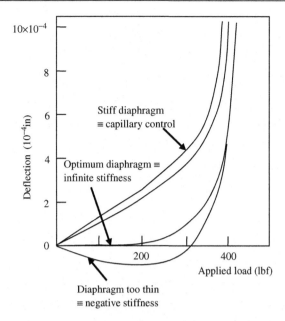

Figure 5.13: Experimentally Derived Characteristics of Diaphragm-Controlled Plane Pad Bearing.

The pressure P_r, from the top reference face, is supplied to what is basically a spool relay valve. The output of the relay feeds an equal pressure P_r to the top face of the main supporting member.

The same principle can be applied to other pressure-sensing valves and an added improvement is the substitution of a diaphragm relay for the spool relay.

The relay valve is required because the slave bearing is a leaking bearing. The need for a relay is avoided if the slave bearing is made into a sealed slave bearing, as in the following example.

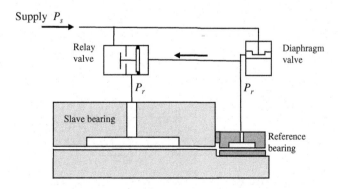

Figure 5.14: Mohsin Reference Bearing System.

Rowe Reference Bearing System

Alternatives to the Mohsin reference bearing system are available for bearings of both single and opposed-pad configurations using the Rowe valve (Rowe and O'Donoghue, 1969–70). An arrangement is illustrated in Figure 5.15, which was designed for correcting straightness errors in long machine tool slideways. This arrangement separates the control of the motion from the effect of errors in the surface supporting the slave bearing. A leaking slave bearing can be employed in the same manner as in the previous Mohsin system by employing a relay between the reference bearing and the slave bearing. Another system that provides film thickness adjustment from a probe was designed by Wong (1965).

Inherent Flow Control

Inherent flow control of hydrostatic pads is sometimes known as "shallow recess control". It differs from other types of flow control in that control is not external to the bearing pad, but depends on its geometry (O'Donoghue and Hooke, 1968–69). A number of geometries have been considered, but the simplest is the circular pad with a shallow recess. With a deep recess, inherent flow control becomes negligible and an external flow control device is necessary.

Inherent flow control based on a shallow recess is illustrated with exaggerated clearances in Figure 5.16, together with typical pressure distributions.

Flow is admitted into the shallow recess through an opening of radius R_1. The pressure reduces as the lubricant flows outwards to radius R_2 at the outer boundary of the shallow recess. The pressure then drops more rapidly as the lubricant flows outward through the bearing land to radius R_3. If the pressure is P_1 at R_1 and P_2 at R_2, the flow in and out at R_2 is given by

$$q_{in} = \frac{\pi(h + h_r)^3(P_1 - P_2)}{6\eta \log_e(R_2/R_1)}$$

$$q_{out} = \frac{\pi h^3 P_2}{6\eta \log_e(R_3/R_2)}$$

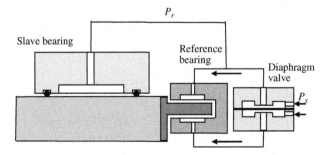

Figure 5.15: Rowe Reference Bearing System for Correcting Errors in Long Machine Tool Slideways.

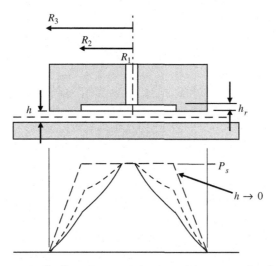

Figure 5.16: Shallow-Recess Inherent Flow Control Principle.

where h is the bearing gap and h_r is the depth of the shallow recess in the bearing. Equating flow in to flow out in the above expressions:

$$\frac{P_2}{P_1} = \frac{1}{1 + \dfrac{\log_e(R_2/R_1)}{\log_e(R_3/R_2)} \cdot \dfrac{1}{(1 + h_r/h)^3}} \tag{5.9}$$

The thrust can be derived in terms of P_1, which is a constant supply pressure, and pressure P_2 from equation (5.9):

$$W = \frac{\pi(R_2^2 - R_1^2)(P_1 - P_2)}{2\log_e(R_2/R_1)} + \frac{\pi(R_3^2 - R_2^2)}{2\log_e(R_3/R_2)} \tag{5.10}$$

Equations (5.1) and (5.2) allow variations in bearing film thickness h with applied load W to be determined.

Yates Combined Journal and Thrust Bearing

The Yates combined journal and thrust hydrostatic bearing is convenient because separate flow controls do not have to be provided for the thrust bearings. The journal bearing acts as a flow control for the thrust bearings; otherwise, one flow control would be required for each thrust pad (Yates, 1950). The Yates bearing is illustrated in Figure 5.17. The shaft arrangement as shown has to be fabricated to allow the bearing to be assembled. This is conveniently arranged by locating one thrust flange on a smaller shaft diameter.

In a combined arrangement, the supply pressure required to support given radial and axial thrust loads for a specific bearing area will be higher than it would be if the bearings were

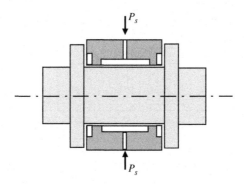

Figure 5.17: Yates Combined Journal and Thrust Hydrostatic Bearing.

separately controlled. Care not to overload the bearing must be taken where loadings in both directions are carried simultaneously.

The Yates combined journal and thrust arrangement has two advantages. It provides an economical arrangement since the supply of lubricant to the thrust bearings is the efflux from the journal bearings. It is also a more compact design of thrust bearing than the conventionally supplied thrust bearing. Surprisingly high stiffness can be achieved in the thrust direction with only small increases in supply pressure.

5.6 General Stiffness Laws for Hydrostatic and Aerostatic Bearing Pads

Bearing stiffness is defined as $\lambda = -\mathrm{d}W/\mathrm{d}h$, where pad thrust is given by $W = P_r A_e$. The stiffness is therefore:

$$\lambda = -\frac{\partial W}{\partial P_r} \cdot \frac{\mathrm{d}P_r}{\mathrm{d}h} - \frac{\partial W}{\partial A_e} \cdot \frac{\mathrm{d}A_e}{\mathrm{d}h}$$

In other words, bearing stiffness consists of two parts, the first dependent only on changes in recess pressure and the second only on changes in the effective pad area:

$$\lambda = -A_e \frac{\mathrm{d}P_r}{\mathrm{d}h} - P_r \frac{\mathrm{d}A_e}{\mathrm{d}h} \tag{5.11}$$

The effective area A_e is constant for most flat pads but varies for nonflat bearings and shallow-recess inherently controlled pads.

Recess pressure P_r is a function of q, h, and $\overline{B}$, since

$$q = P_r \frac{\overline{B}h^3}{\eta} \qquad \text{Liquid}$$

$$q = \frac{p_r^2 - p_a^2}{2p_a} \frac{\overline{B}h^3}{\eta} \qquad \text{Gas} \tag{5.12}$$

Equation (5.11) is therefore expanded, giving

$$\lambda = -A_e \frac{dP_r}{dq} \left[\frac{\partial q}{\partial h} + \frac{\partial q}{\partial \overline{B}} \frac{d\overline{B}}{dh} + \frac{\partial q}{\partial P_r} \frac{dP_r}{dh} \right] - P_r \frac{dA_e}{dh}$$

The partial derivatives for liquid flow are obtained from equations (5.12) so that

$$\lambda = -A_e \frac{dP_r}{dq} \left[\frac{3q}{h} + \frac{q}{\overline{B}} \frac{d\overline{B}}{dh} + \frac{q}{P_r} \frac{dP_r}{dh} \right] - P_r \frac{dA_e}{dh}$$

With simplification, general stiffness laws are found:

$$\frac{\lambda}{W} = -\frac{\dfrac{3}{h} + \dfrac{1}{\overline{B}} \dfrac{d\overline{B}}{dh}}{1 - \dfrac{P_r}{q} \dfrac{dq}{dP_r}} - \frac{1}{A_e} \frac{dA_e}{dh} \qquad \text{Liquid}$$

$$\frac{\lambda}{W} = -\frac{\dfrac{3}{h} + \dfrac{1}{\overline{B}} \dfrac{d\overline{B}}{dh}}{\dfrac{2p_r}{p_r + p_a} - \dfrac{p_r - p_a}{q} \dfrac{dq}{dP_r}} - \frac{1}{A_e} \frac{dA_e}{dh} \qquad \text{Gas}$$

(5.13)

Equations (5.13) enable stiffness to be determined for a bearing having combined inherent and external flow control. In most cases, only one method of control applies.

Stiffness of Inherently Compensated Bearings

In the case of an inherently compensated bearing with constant recess pressure, only the last term in equation (5.13) is relevant, so that

$$\lambda = P_r \frac{dA_e}{dh}$$

where dA_e/dh is obtained from equations such as (5.1) and (5.2).

Stiffness of Externally Compensated Plane Pads

For purely external flow control, pad shape factors are constant, so that the first term of equation (5.13) for hydrostatic pads applies and $d\overline{B}/dh$ is zero:

$$\lambda = \frac{3W}{h} \cdot \frac{1}{1 - \dfrac{P_r}{q} \dfrac{dq}{dp_r}} \qquad (5.14)$$

Inspection of equation (5.14) reveals that there are three methods of increasing the stiffness of the bearing. The first is to increase the operating thrust W. The second is to reduce the operating film thickness h. The third is to reduce the value of the denominator of the remaining term.

In an externally controlled hydrostatic bearing, the value of this term depends only on the characteristics of the flow control device. Thus,

If $\dfrac{P_r}{q}\dfrac{dq}{dp_r} = 1$, the bearing has infinite static stiffness

If $\dfrac{P_r}{q}\dfrac{dq}{dp_r} < 1$, the bearing is stable and stiffness is positive

If $\dfrac{P_r}{q}\dfrac{dq}{dp_r} > 1$, the bearing is statically unstable since stiffness is negative

The four main types of control are compared in Figure 5.4. Capillary and simple laminar-flow restrictors give the lowest stiffness but tolerate temperature variations better than other types of control. Orifice restrictors give higher stiffness than capillary restrictors at the design condition and constant flow even higher, but orifices are affected by changes in viscosity. A pressure-sensing valve tuned to give a flow-pressure slope of $+1$ gives infinite static stiffness. In practice, it is usual to design for lower stiffness than infinite to ensure stability and avoid the possibility of limit-cycle oscillations. Valves usually exhibit nonlinearity across the load range so that stiffness is reduced as maximum pressure is approached. This applies both to constant-flow valves and pressure-sensing valves.

5.7 Aerostatic Bearings and Flow Control

The system to supply an aerostatic air bearing is often much simpler than the system required for a hydrostatic bearing. Air at gauge pressure up to 6 bars is commonly available in the workshop and can be employed with little extra equipment. The main requirements are for careful filtration, moisture traps, a pressure regulator, an accumulator in case of supply pressure failure, and a pressure relay switch that cuts off drive motors to prevent material damage (Figure 5.18).

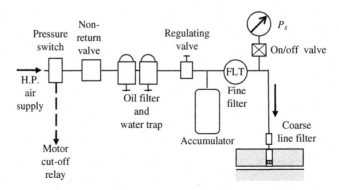

Figure 5.18: A Supply Circuit for an Aerostatic Bearing.

The principles outlined in Section 5.5 apply in broad outline for aerostatic bearings, although detail design differs and restrictors are designed employing the correct form for compressible flow, as outlined in Chapter 2.

Aerostatic Orifice Restrictor and Effective Restriction Area

Orifice restrictors were introduced in Section 5.5. The characteristics of aerostatic orifice control require further consideration as follows. A typical orifice restrictor for aerostatic bearings is shown in Figure 5.19. The basic orifice has a restriction area $\pi d_o^2/4$. This restriction area must be less than the secondary annular restriction area $\pi d_r h$ formed by the entrance to the bearing film from the recess.

A typical diameter of a basic orifice is 0.1–0.4 mm. The recess diameter is typically 2–3 mm, and the depth approximately 0.2–0.4 mm. The requirement is that $d_o^2 < 4d_r h$. For example, if $d_o = 0.3$ mm, $d_r = 2$ mm, and $h = 0.05$ mm, $d_o^2 = 0.09$ mm^2 and $4d_r h = 0.4$ mm^2, so that the annular restriction area is 4.44 times the basic orifice restriction area. The dominant restriction will therefore be formed by the basic orifice.

The recess depth of an aerostatic bearing is usually very small and it is therefore possible for a further annular restriction to be formed around the curtain at the entrance from the orifice into the recess. This curtain area entering the recess is given by $\pi d_o(h + h_r)$. MTI (Wilcock, 1967) combined the restriction values into one expression by employing a restrictor ratio:

$$\delta = \sqrt{\left(\frac{d_o^2}{4d_r h}\right)^2 + \left(\frac{d_o}{4(h + h_r)}\right)^2}$$

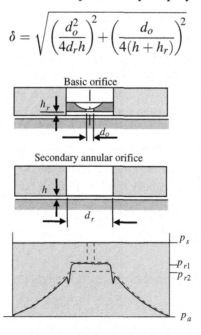

Figure 5.19: Aerostatic Orifice Restrictor and Typical Pressure Distribution Showing Effect of Secondary Entrance Pressure Losses.

A single orifice area can then be employed by writing the orifice restriction area as

$$A_o = \frac{\pi d_o^2}{4\sqrt{1 + \delta^2}} \tag{5.15}$$

For basic orifice control, Pink and Stout (1978) recommended $\delta \leq 0.5$. Other evidence suggests that the recess only needs to be deep enough to ensure $\delta \leq 0.7$ (Grewal, 1979).

When δ becomes large, the equation is identical to the expression for an annular orifice but as δ decreases, the expression becomes progressively equal to the value for a basic orifice. In deducing this expression, it was assumed that C_d for an annular orifice is approximately equal to C_d for a basic orifice.

The design and manufacture of orifice restrictors is made more challenging by the requirement to ensure a very small recess volume and by the need to drill small holes avoiding burrs. Orifices have been successfully manufactured from brass using a similar design to that shown in Figure 5.7. The hole in which the nozzle is located must be accurately reamed and the nozzle itself should be a close fit within the body to prevent an air volume surrounding the nozzle body.

A common practice is to machine and drill ruby gems to the required dimensions. These may be pressed into a hole of the required recess diameter. Detailed guidance to the dimensional design of orifice-compensated aerostatic journal bearings was given by Pink and Stout (1978).

Figure 5.19 shows entrance effects on pressure distribution where the basic orifice restrictor is not completely dominant. The pressure in the recess at the exit from the orifice drops to the stagnation pressure p_{r1}. On entry to the bearing film, the annular restriction at the edge of the recess forms a secondary resistance. The pressure initially drops more than expected as velocity builds up. But as viscous resistance becomes dominant, pressure recovers again. The pressure recovery is less complete for radial flow pads than for linear flow bearings and also at high supply pressures. In a well-designed pad, with recess pressures less than $p_r = 3$ bar, entry effects on pressure distribution are usually negligible but further information on models of entry losses are given by Gross (1962), who also gives relationships for turbulent flow in the bearing film. Air flow, although turbulent in an orifice, is usually nonturbulent in the bearing film.

The restriction offered by an annular restrictor increases as the bearing film thickness decreases. This reduces bearing stiffness by one-third, as is the case for hydrostatic bearings, but reduces the risk of instability.

Aerostatic bearings with orifice control are usually designed to ensure unchoked orifice flow. Orifices become choked when

$$\frac{P_r}{P_s} \leq \left[\frac{2}{k+1}\right]^{\frac{k}{k-1}}$$

Choked flow increases the risk of instability evidenced by mechanical vibration.

For air, the transition to choked flow is given by $p_r/p_s = 0.528$ and it is necessary to check the selected pressure ratio for the choked condition. For example, if $K_{go} = 0.5$ and $p_a/p_s = 0.2$, $p_r/p_s = 0.6$ and therefore flow is unchoked.

Aerostatic Slot and Laminar-Flow Restrictors

Figure 5.20 shows a slot restrictor and a capillary restrictor feeding into a slot-shaped groove in the pad surface. Both cases show simplified pressures for longitudinal pad flow. The simplified pressures are identical for the two cases. The mass flow expressions assume that gas density is stated at ambient pressure.

A capillary restrictor does not need to feed into a slot-shaped groove any more than an orifice restrictor does. Figure 5.21 illustrates a longitudinal flow situation employing a row of holes

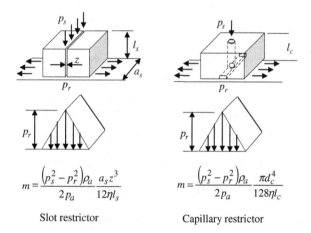

$$m = \frac{\left(p_s^2 - p_r^2\right)\rho_a}{2p_a} \frac{a_s z^3}{12\eta l_s}$$

Slot restrictor

$$m = \frac{\left(p_s^2 - p_r^2\right)\rho_a}{2p_a} \frac{\pi d_c^4}{128\eta l_c}$$

Capillary restrictor

Figure 5.20: An Aerostatic Slot Restrictor and a Capillary Restrictor for Slot-Entry Feeding Conditions Showing Pressures in Longitudinal Flow.

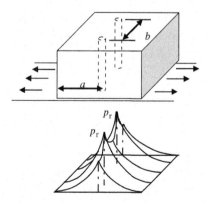

Figure 5.21: An Aerostatic Hole-Entry Pad for Linear Flow and Effect of Hole Separation on Dispersion Losses of Pressure.

instead of a groove. With wide separation between the holes, there is a pressure drop between the restrictors (see also Figure 4.1). This is known as the "dispersion effect" because flow from the holes has to flow sideways to fill the gap. The consequence of hole-entry feed into a bearing and the need for dispersion is a loss of aerostatic load support. Pressure drop between the holes can be minimized by reducing the distance b between the holes. The ratio b/a should be less than 0.5. For example, a journal bearing with central admission, having $L/D = 1$, requires more than six holes per row. In practice, 12 holes per row is much better.

Instability and Aerostatic Recess Volume

Gross (1962) recommended that recess volume should be minimal to reduce the risk of instability. Instability may be evidenced as a self-excited vibration or "pneumatic hammer" that can be felt through the bearing structure. Gross recommended that recess depth should be of the same order as the bearing film thickness. Both should be minimized for stability. A high pressure ratio K_{go} should be employed and orifice diameter should be large. He also comments that reducing bearing stiffness is helpful. It was further suggested that increasing supply pressure reduces stability and increasing recess pressure improves stability. Powell (1970) recommends that total hole and recess volume should be less than 5% of the total bearing film volume.

An alternative way of considering the problem is to consider the recess volume in relation to the bearing flow. For example, consider a case where the recess dimensions are 3 mm diameter by 0.4 mm depth, so that the recess volume is 2.8 mm^3. If volume flow has the value 2×10^3 mm^3/s, it means that recess volume divided by flow is 1.4 milliseconds. In other words, the recess volume is capable of being filled at a frequency of 714 Hz. This appears satisfactory since few machines suffer severe vibrations at such a high frequency. However, it would be easy, with a loose-fitting nozzle and a deeper recess, for the ratio to be one-tenth of the previous value so that the response frequency is reduced to 71 Hz. Such a low value might appear very much less satisfactory. While there is no experimental validation for this approach, the logic agrees with risk of pneumatic hammer increasing with recess volume and also gives qualitative agreement with the findings of Gross. It is tentatively suggested that a design rule be applied as follows for air flow:

$$f = \frac{\text{orifice flow}}{\text{recess volume}} \geq 1000 \text{ Hz} \qquad (5.16)$$

There is some additional evidence of an alternative method for preventing or overcoming instability. It appears that connecting a small gas chamber to the recess through a restrictor can sometimes damp out vibrations. This provides a possible alternative method of overcoming a particular problem, but no details of a design methodology are available.

Incorporating a central land within a recess area creates a virtual recess, as explained in the previous chapter. Substituting a virtual recess for a real recess increases the squeeze film area while allowing the entry source volume to be minimized and improves stability.

References

Grewal, S. S. (1979). *An investigation of externally-pressurised orifice-compensated air journal bearings with particular reference to mis-alignment and inter-orifice variations.* CNAA PhD thesis. Liverpool Polytechnic.

Gross, W. A. (1962). *Gas Film Lubrication.* New York: John Wiley.

Mohsin, M. E. (1962). *The use of controlled restrictors for compensating hydrostatic bearings. Advances in Machine Tool Design and Research.* Oxford: Pergamon Press.

O'Donoghue, J. P., & Hooke, C. J. (1968–69). Design of inherently stable hydrostatic bearings. Tribology Convention, *Proceedings of the Institution of Mechanical Engineers*, London, Gothenburg. *Sweden, Vol. 183*(Part 3P), 172–176.

Pink, E. G., & Stout, K. J. (1978). Design procedure for orifice compensated gas journal bearings based on experimental data. *Tribology International, February* 63–75.

Powell, J. W. (1970). *The Design of Aerostatic Bearings.* UK: Machinery Publishing Company.

Rippel, H. C. (1963). Design of hydrostatic bearings, Part 3. Influence of restrictors. *Machine Design, 29 August*, 132–138.

Rowe, W. B. (1966). *Hydrostatic Bearings.* UK Patent 1 170 602, November 1969 (application 22072/66, 18 May 1966).

Rowe, W. B., & O'Donoghue, J. P. (1969–70). Diaphragm valves for controlling opposed pad bearings. Tribology Convention, Brighton. *Proceedings of the Institution of Mechanical Engineers, London, Vol. 184*(Part 3L), 1–9.

Royle, J. K., et al. (1962). Applications of automatic control to pressurized oil-film bearings. *Proceedings of the Institution of Mechanical Engineers, London, Vol. 176*(22), 532–541.

Wilcock, D. F. (Ed.). (1967). *Design of Gas Bearings.* LA: Mechanical Technology Incorporated.

Wong, G. S. K. (1965). Interface restrictor hydrostatic bearings. *Proceedings of the 6th International Machine Tool Design and Research Conference.* Manchester University. Oxford: Pergamon Press.

Yates, S. (1950). *Combined journal and thrust bearings.* UK Patent 639 293, June.

Appendix: Tabular Design Procedures for Restrictors and Worked Examples

Procedure 5.A1 Capillary Restrictor for a Hydrostatic Bearing

Step	Symbol	Description of Operation	Notes
1	K_c	Calculate: capillary factor	From $K_c = (1-\beta)P_s/q_r\eta$
2	l_c/d_c	Specify: length/diameter	Suggestion: > 20
3	d_c	Calculate: capillary diameter	From Figure 5.3 or $d_c = \sqrt[3]{\dfrac{128}{\pi K_c}\cdot\dfrac{l_c}{d_c}}$
4	l_c	Calculate: length	From $l_c = d_c \cdot l_c/d_c$
5	Re	Calculate: Reynolds number	From $Re = \dfrac{4\rho q_r}{\pi d_c \eta}$

Procedure 5.A2 Slot Restrictor for a Hydrostatic Bearing

Step	Symbol	Description of Operation	Notes
1	K_s	Calculate: slot factor	From $K_s = (1-\beta)P_s/q_r\eta$
2	l_s/a_s	Specify: length/width	Suggestion: > 0.5
3	a_s	Specify: slot width	Surround at least 75% of a virtual recess
4	l_s	Calculate: length	From $l_s = a_s \cdot l_s/a_s$
5	z_s	Calculate: restrictor thickness	From $z_s = \sqrt[3]{\dfrac{12\,l_s}{K_s\,a_s}}$
6	Re	Calculate: Reynolds number	From $Re = \dfrac{\rho q_r}{a_s\eta}$

Procedure 5.A3 Slot Restrictor for an Aerostatic Bearing

Step	Symbol	Description of Operation	Notes
1	K_s	Calculate: slot factor	From $K_s = \dfrac{(p_s^2 - p_{ro}^2)}{2p_a q_r \eta}$
2	l_s/a_s	Specify: length/diameter	Suggestion: > 0.5
3	a_s	Specify: slot width	Surround at least 75% of a virtual recess
4	l_s	Calculate: slot length	From $l_s = a_s \cdot l_s/a_s$
5	z_s	Calculate: restrictor thickness	From $z_s = \sqrt[3]{\dfrac{12\,l_s}{K_s\,a_s}}$
6	Re	Calculate: Reynolds number	From $Re = \dfrac{\rho q_r}{a_s\eta}$

Example 5.A1 Hydrostatic Capillary Restrictor: $P_s = 2$ MN/m^2, $\beta = 0.5$, Oil Viscosity $\eta = 0.035$ N s/m^2, Density $\rho = 870$ kg/m^3, $q_r = 2.5 \times 10^{-6}$ m^3/s

Step	Symbol	Example of Working	Result
1	K_c	$K_c = (1 - 0.5) \times 2 \times 10^6 / 2.5 \times 10^{-6} \times 0.035$	11.43×10^{12} m^{-3}
2	l_c/d_c	Length/diameter	200
3	d_c	$\sqrt[3]{\dfrac{128}{\pi} \times \dfrac{200}{11.43 \times 10^{12}}}$	0.00089 m (0.0352 in)
4	l_c	$l_c = 0.00089 \times 200$	0.178 m (7 in)
5	Re	$\dfrac{4 \times 870 \times 2.5 \times 10^{-6}}{\pi \times 0.00089 \times 0.035}$	89, i.e. laminar

Example 5.A2 Hydrostatic Slot Restrictor: $P_s = 2$ MN/m^2, $\beta = 0.5$, Oil Viscosity $\eta = 0.035$ N s/m^2, Density $\rho = 870$ kg/m^3, $q_s = 0.25 \times 10^{-6}$ m^3/s for a 2×12 Slot/Row Journal Bearing of 50 mm Diameter

Step	Symbol	Example of Working	Result
1	K_s	$K_s = (1 - 0.5) \times 2 \times 10^6 / 0.25 \times 10^{-6} \times 0.035$	114.3×10^{12} m^{-3}
2	l_s/a_s	Length/diameter	0.5
3	a_s	$0.85 \times \pi \times 0.05/12$	0.01113 m (0.438 in)
4	l_s	0.01113×0.5	0.005565 m (0.2224 in)
5	z_s	$\sqrt[3]{\dfrac{12}{114.3 \times 10^{12}}} \times 0.5$	0.00003744 m (0.00147 in)
6	Re	$\dfrac{870 \times 0.25 \times 10^{-6}}{0.1113 \times 0.035}$	0.056, i.e. laminar

Example 5.A3 Aerostatic Slot Restrictor: $p_a = 0.101$ MN/m^2, $P_s = 0.414$ MN/m^2, $p_s = 0.515$ MN/m^2, $K_{go} = 0.5$, $p_{ro} = 0.308$ MN/m^2, Air Viscosity $\eta = 0.0000182$ N s/m^2, $\rho = 1.208$ kg/m^3 and $q_r = 60 \times 10^{-6}$ m^3/s Free Air

Step	Symbol	Example of Working	Result
1	K_s	$\dfrac{(0.515^2 - 0.308^2) \times 10^{12}}{2 \times 0.101 \times 10^6 \times 60 \times 10^{-6} \times 0.0000182}$	772.3×10^{12} m^{-3}
2	l_s/a_s	Length/diameter	0.5
3	a_s		0.015 m (0.591 in)
4	l_s	$l_s = 0.015 \times 0.5$	0.0075 m (0.295 in)
5	z_s	$\sqrt[3]{\dfrac{12}{772.3 \times 10^{12}}} \times 0.5$	0.0000198 m (0.00078 in)
6	Re	$\dfrac{1.208 \times 60 \times 10^{-6}}{0.015 \times 0.0000182}$	265, i.e. laminar

Procedure 5.A4 Hydrostatic Orifice Restrictor

Step	Symbol	Description of Operation	Notes
1	C_d	Specify: orifice factor	From Figure 2.7. First estimate $C_d = 0.6$
2	A_o	Calculate: orifice area	$A = \dfrac{q_r}{C_d} \sqrt{\dfrac{\rho}{2(P_s - P_{ro})}}$
3	d_o	Calculate: orifice diameter	From $d_o = \sqrt{\dfrac{4A_o}{\pi}}$
4	Re	Calculate: Reynolds number	From $Re = \dfrac{4\rho q_r}{\pi d_o \eta}$
5	C_d	Specify: second estimate	From Figure 2.7

Procedure 5.A5 Aerostatic Orifice Restrictor

Step	Symbol	Description of Operation	Notes
1	K_p	Calculate: pressure ratio	p_{ro}/p_s
2	C_d	Specify: orifice factor	From Figure 2.7
3	C_{kp}	Calculate or read from Figure 2.8	$C_{kp} = \sqrt{\dfrac{k}{k-1}\left[K_p^{2/k} - K_p^{(1+1/k)} \right]}$
4	A_o	Calculate: orifice area	$A_o = \dfrac{q_r}{C_d C_{kp}} \dfrac{p_a}{p_s} \sqrt{\dfrac{\rho_a}{2p_a}}$
5	d_o	Calculate: orifice diameter	From $d_o = \sqrt{\dfrac{4A_o}{\pi}}$

Example 5.A4 Hydrostatic Orifice Restrictor: $P_s = 2$ MN/m^2, $\beta = 0.5$, Oil Viscosity $\eta = 0.035$ N s/m^2 and Density $\rho = 870$ kg/m^3, $q_s = 2.5 \times 10^{-6}$ m^3/s

Step	Symbol	Example of Working	Result
1	C_d	Orifice factor	0.65 (see step 5)
2	A_o	$\dfrac{2.5 \times 10^{-6}}{0.65}\sqrt{\dfrac{870}{2 \times (2-1) \times 10^6}}$	8×10^{-8} m^2 (1.24×10^{-4} in)
3	d_o	$\sqrt{4 \times 8 \times 10^{-8}/\pi}$	0.0003192 m (0.0126 in)
4	Re	$\dfrac{4 \times 870 \times 2.5 \times 10^{-6}}{\pi \times 0.0003192 \times 0.035}$	248, i.e. turbulent for short jet
5	C_d	Second estimate	From Figure 2.7. $C_d = 0.65$

Example 5.A5 Aerostatic Orifice Restrictor: $p_a = 0.101$ MN/m^2, $P_s = 0.414$ MN/m^2, $p_s = 0.515$ MN/m^2, $K_{go} = 0.5$, $p_{ro} = 0.308$ MN/m^2, Air Viscosity $\eta = 0.0000182$ N s/m^2, $\rho = 1.208$ kg/m^3 and $q_r = 60 \times 10^{-6}$ m^3/s Free Air; For Air $k = 1.41$

Step	Symbol	Example of Working	Result
1	K_p	0.308/0.515	0.5981, i.e. unchoked flow
2	C_d	Orifice factor	0.85
3	C_{kp}	$C_{kp} = \sqrt{\dfrac{1.41}{1.41-1}\left[0.5981^{2/1.41} - 0.5981^{(1+1/1.41)}\right]}$	0.48
4	A_o	$\dfrac{60 \times 10^{-6}}{0.85 \times 0.48} \times \dfrac{0.101}{0.515} \times \sqrt{\dfrac{1.208}{2 \times 0.101 \times 10^6}}$	7.053×10^{-8} m^2 (1.093×10^{-4} in)
5	d_o	$\sqrt{4 \times 7.053 \times 10^{-8}/\pi}$	0.0003 m (0.0118 in)

CHAPTER 6

Basis of Design Procedures

6.1 Introduction

For single-pad bearings, film thickness depends on the applied load and may therefore be considered to be self-adjusting. This means either supply pressure or pad size can be adjusted to achieve optimum pressure ratio at any applied load.

For opposed pads and journals, bearing clearances must be specified. Clearance tolerances are usually required. A basis for selection of tolerances is suggested in this chapter.

For very-low-speed bearings, design is simple. For high-speed bearings, it is recommended that power ratio is specified since power ratio has implications for many aspects of performance.

In all cases, design procedures involve two main parts:

- Part I: Design of bearing geometry for load, flow, and stiffness
- Part II: Design of restrictors for flow control based on flow from Part I.

6.2 An Acceptable Range for Design Pressure Ratio

Generally, a pressure ratio of 0.5 is approximately optimum for both hydrostatic and aerostatic bearings because this value allows the maximum load range.

A tolerance range for bearing clearance may be selected for opposed-pad bearings so that design pressure ratio lies within a permissible range for restrictors of fixed resistance. This produces a range for pressure ratio from 0.4 to 0.7 when maximum and minimum clearance limits are in the proportion 1.5:1. The selection of these tolerances means that restrictors manufactured to produce maximum pressure ratio at minimum clearance will then result in minimum pressure ratio at maximum clearance.

An alternative approach that overcomes deviations from set tolerances is to specify that flow restrictors are manufactured to suit measured clearance. This allows restrictors to be manufactured to achieve a design pressure ratio very close to 0.5 or any other value required.

For orifice-fed aerostatic bearings, low pressure ratios increase the risk of pneumatic hammer instability. Minimum recommended pressure ratio for orifice feed is 0.5.

Hydrostatic, Aerostatic and Hybrid Bearing Design.
DOI: 10.1016/B978-0-12-396994-1.00006-1

6.3 Zero- and Low-Speed Hydrostatic Bearings

Design of low-speed bearings is straightforward, as indicated in Figure 6.1. A low-speed bearing is where speed, length, film thickness, and viscosity lead to power ratio $K = H_f/H_p < 1$. Power ratio K is a measure of bearing speed that takes all bearing parameters into account.

"Clearance" is a term used to describe the manufactured gap between a rotor or slider and a bearing stator. For journal bearings, radial clearance h_o is the radial gap between the journal and the bearing when the journal is concentric within the bearing. Diametral clearance is twice radial clearance.

The film thickness of a single plane pad depends on the applied load on the bearing, the supply pressure and the restrictor resistance. In this case, a designer will aim for the smallest film thickness compatible with surface geometry and accuracy, select values of P_s and design pressure ratio β or K_{go}, and design the flow control device accordingly.

Design procedures in later chapters are based on the stages indicated in Figure 6.1:

1. Choose bearing type, for example, opposed rectangular pad bearing.
2. Determine parameters constrained by machine design considerations. For example,

    Load, W
    Size, L and B
    Speed, $N = 0$.

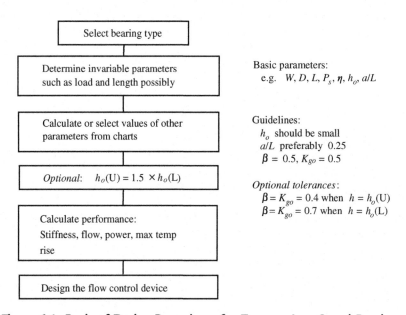

Figure 6.1: Basis of Design Procedures for Zero- or Low-Speed Bearings.

3. Determine parameters where these may be selected. For example,

   Land-width ratio, a/L

   Pressure ratio, β

   Clearance (design film thickness), h_o

   Viscosity, η.

4. Determine parameters that must be calculated. For example,

   Supply pressure $P_s = W/A\overline{A}\,\overline{W}$ for thrust pads or

   Supply pressure $P_s = W/LD\overline{W}$ for journals

   $\overline{A}$ and $\overline{W}$ taken from charts or formulae

   Optional: clearance upper and lower limits, $h_o(\text{U})$ and $h_o(\text{L})$

   For example, $h_o(\text{U}) = 1.5 \times h_o(\text{L})$.

5. Calculate performance data from charts or formulae. For example,

   Bearing film stiffness, λ

   Flow, q

   Pumping power, H_p

   Friction power, H_f

   Total power, H_t

   Maximum temperature rise for one oil pass, ΔT.

6. Design the flow control device for flow and pressure ratio. For example, capillary dimensions,

   Capillary diameter, d_c

   Capillary length, l_c.

Design procedures in later chapters are combined for low-speed and high-speed bearings so that it is unnecessary to determine in advance whether a bearing is low speed or high speed.

6.4 Zero- and Low-Speed Aerostatic Bearings

The design of a low-speed aerostatic bearing is usually straightforward and a typical procedure is indicated in Figure 6.1. Most aerostatic bearings are low speed, where basic parameters such as speed, length, clearance, and viscosity lead to power ratio $K = H_f/H_p \leq 1$.

For orifice-fed aerostatic bearings, it is recommended that the design pressure ratio does not lead to choked jets. This requires that low pressure ratios should be avoided for orifice-fed bearings.

Smallest film thickness should be selected compatible with surface geometry and accuracy. Supply pressure P_s is selected to support the load. Gauge pressure ratio is selected as above. Flow through the bearing and flow through the restrictors can then be evaluated, and flow restrictors are designed accordingly.

Procedures are given in appropriate chapters with worked examples.

6.5 Optional Size Limits and Tolerances for Film Thickness

ISO definitions of size limits and tolerances are described later in this chapter. Size limits consist of an upper limit and a lower limit. The tolerance is the difference between the upper and lower limits. For fluid-film bearings, film thickness between mating parts is critical. Restrictors are a relatively low-cost item and can be adjusted to meet tight tolerances. Where so desired, this allows the main tolerance to be allocated to bearing clearance. A basis for selecting clearance tolerances is described as follows.

A lower clearance limit $h_o(L)$ to be employed with pressure ratio equal to $\beta = 0.7$ is set for hydrostatic bearings or $K_{go} = 0.7$ for aerostatic bearings. It is found that pressure ratio for hydrostatic bearings drops to 0.409 when the upper clearance limit $h_o(U)$ is set 50% higher than the lower clearance limit. It is shown in the following chapters for the main applications that the range $0.4 < \beta < 0.7$ spans the high-load range. Higher or lower values lead to significant loss of load support.

The reader may check the validity of the above finding from equations (5.1) and (5.3), leading to the following expression for pressure ratio. Starting from $\beta = 0.7$ when $\overline{X} = 1$, pressure ratio when $\overline{X} = 1.5$ is given by

$$\frac{P_r}{P_s} = \frac{1}{1 + \dfrac{1-\beta}{\beta}\overline{X}^3} = \frac{1}{1 + \dfrac{1-0.7}{0.7} \times 1.5^3} = 0.409$$

For orifice-fed aerostatic bearings, a reduced range may be employed where $0.5 < K_{go} < 0.7$. A reduced range reduces the risk of pneumatic instability. In this case, $h_o(U) \approx 1.25 h_o(L)$ for orifice-fed aerostatic bearings.

Variation in manufactured clearance from the specified value changes the power ratio K for bearings that operate at speed. This happens because an increase in clearance increases flow, according to equation (2.23), and hence increases pumping power. At the same time, friction power is reduced according to equation (3.5) and power ratio is reduced.

The combined effects on pressure ratio and power ratio can be evaluated. While pressure ratio drops from 0.7 to 0.4 as clearance increases by 50%, power ratio, if set initially to $K = 3$, falls to $K = 1$. This is in accordance with the recommended range $1 < K < 3$. This range yields maximum stiffness and minimum power.

The clearance range applies equally well to low-speed and high-speed bearings. Where a design procedure does not make specific reference to tolerances, a designer may achieve tolerances, if so desired, by designing a bearing based on the upper limit of clearance and the associated values for pressure ratio and power ratio, as given in Table 6.1. The lower limit of clearance is then set to two-thirds of the upper limit.

Table 6.1

Upper Clearance Limit	Lower Clearance Limit
$\beta = K_{go} = 0.7$	$\beta = K_{go} = 0.4$
$h_o(L)$	$h_o(U) = 1.5 \times h_o(L)$
$K = 1$	$K = 3$

A reduced tolerance range may be applied for orifice-fed aerostatic bearings, as described above, to reduce the risk of pneumatic instability. If problems are experienced with pneumatic hammer, it is always possible to experiment with slightly larger orifices, which may sometimes cure a problem by increasing pressure ratio.

The importance of correct selection of tolerances should not be underestimated. If the designer specifies a very small tolerance, cost is increased and possibly the specification is not achieved.

■ *Example 6.1 Selection of Clearance Limits*

Bearing clearance limits are 0.025 mm (0.001 in) and 0.030 mm (0.0012 in), so that the tolerance is 0.005 mm (0.0002 in). Will performance be satisfactory across the range of possible sizes?

Solution

If the associated parameters are chosen so that $\beta = 0.4$ when $h_o = 0.030$ mm, the designer may rest assured of satisfactory performance, although it will be seen that the tolerance is smaller than necessary since $h_o(U)/h_o(L) = 1.2$ instead of the permitted 1.5.

Recommended values for increased tolerance are:

$h_o(L) = 0.025$ mm when $\beta = 0.7$ or $K_{go} = 0.7$
$h_o(U) = 1.5 \times 0.025 = 0.037$ mm when $\beta = 0.4$ for a hydrostatic bearing
$h_o(U) = 1.25 \times 0.025 = 0.031$ mm when $K_{go} = 0.5$ for an orifice-fed aerostatic bearing.

■

6.6 High-Speed Bearings

The basis of design procedures for high-speed opposed-pad hydrostatic bearings is illustrated in Figure 6.2 and for high-speed hydrostatic journal bearings in Figure 6.3. Minimization of total power is not usually relevant for aerostatic bearings, where pumping power usually exceeds friction power by a large margin.

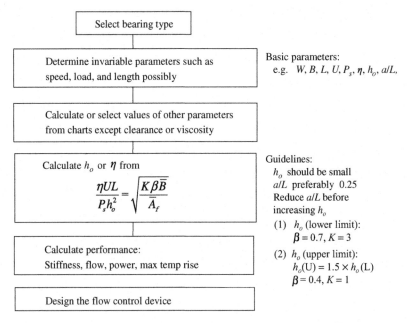

Figure 6.2: Basic Procedure for High-Speed Opposed-Pad Bearings.

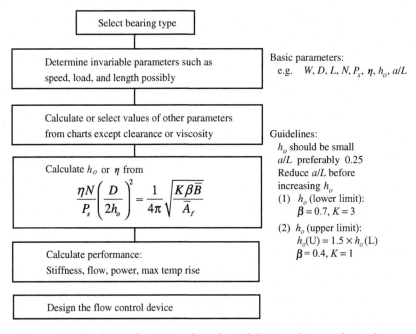

Figure 6.3: Basic Design Procedure for High-Speed Journal Bearings.

Design procedures given in each chapter do not require tolerances to be evaluated, although provision is made for tolerances by employing one of the clearance limits recommended above together with the associated values of pressure ratio and power ratio.

The strategy for high-speed bearings listed above introduces a minor modification from the strategy for low-speed bearings. The selection of clearance is no longer independent of the value of viscosity. The two parameters are linked by the necessity to ensure minimum power. The difference from the strategy of Section 6.1 occurs at stage 4 as follows:

1. Choose bearing type. For example, recessed journal bearing.
2. Determine parameters constrained by machine design considerations. For example,
    Load, W
    Size, D or L
    Speed, N.
3. Determine parameters that may be selected. For example,
    Land-width ratio, a/L
    Inter-recess land-width ratio, b/a
    Length/diameter ratio, L/D
    Lower radial clearance limit, $h_o(\text{L})$
    Pressure ratio, $\beta = 0.7$
    Power ratio, $K = 3$.
4. Determine parameters to be calculated. For example,

$$\text{Supply pressure } P_s = \frac{W}{LD\overline{W}} \text{ reading } \overline{W} \text{ from charts}$$

$$\text{Speed parameter } S_h = \frac{1}{4\pi}\sqrt{\frac{K\beta\overline{B}}{\overline{A}f}} \text{ from chart or by calculation}$$

$$\text{Viscosity } \eta = S_h \Bigg/ \left(\frac{N}{P_s}\left(\frac{D}{2h_o}\right)^2\right) \text{ or for gas or if more appropriate}$$

$$\text{Clearance ratio } \frac{2h_o}{D} = \sqrt{\frac{\eta N}{P_s S_h}}$$

Upper radial clearance limit $h_o(\text{U}) = 1.5 \times h_o(\text{L})$.

5. Determine performance data from charts or formulae. For example,
    Bearing film stiffness, λ
    Flow rate, q
    Pumping power, H_p
    Friction power, H_f
    Total power, H_t
    Maximum temperature rise for one oil pass, ΔT.

6. Design restrictors for flow and pressure ratio. For example,
   Capillary diameter, d_c
   Capillary length, l_c.

6.7 Specification for ISO Limits and Fits

The International Standards Organization (ISO) has not made recommendations for the sizing of fluid-film bearings since any such recommendations must be based on fluid-film theory. In fact, advisory bodies have questioned the desirability of trying to standardize clearances for bearings. The procedure incorporated into the design procedures given in this book are probably as advanced as may be considered practicable or even desirable for some time to come.

The terminology employed by the ISO is used wherever possible in the design procedures. However, an important difference arises. In the ISO system, the shaft and the bearing are each given an upper size limit and a lower size limit. The difference between the upper limit and the lower limit of each part is the size tolerance. The tolerance is related to the size of the part according to the tolerance grade. The hole, for example, is given an upper size limit and a lower size limit, the difference being the hole tolerance, which is also related to the nominal diameter of the hole according to a tolerance grade. These are known as International Tolerance (IT) Grades and are associated with a number indicative of quality, e.g. IT7 is average for precision machining and IT4 is very high quality. Examples of tolerances and International Tolerance Grades from BS4500 are given approximately in Figure 6.4.

6.8 Tolerance Grades for Hydrostatic and Aerostatic Bearings

For hydrostatic and aerostatic bearing design, shaft tolerance and hole tolerance are added and given as one clearance tolerance. The designer can apportion tolerances to the shaft and the hole diameters from knowledge of the machining processes to be employed. Tolerances and proposed diametral clearance limits for hydrostatic and aerostatic bearings are given in Figure 6.5. These are suggestions and there is no compulsion to accept the recommended values.

Rowe and Stout (1975) and Stout and Rowe (1974) give further details on effects of manufacturing errors on performance.

The clearance limits in Figure 6.5 are derived from Figure 6.4 in a simple way. The upper clearance limit is 50% higher than the lower clearance limit, and the difference between the two limits is equal to the combined tolerance on the hole and the shaft diameters. A reduced tolerance range can be achieved if required by specifying $h_o(U) = 1.25h_o(L)$.

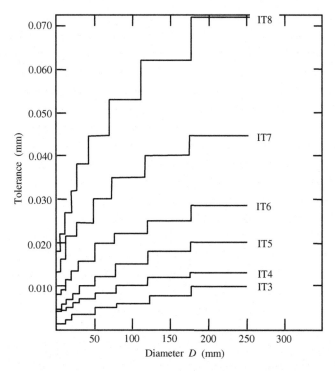

Figure 6.4: International Tolerance Grades and Tolerances.

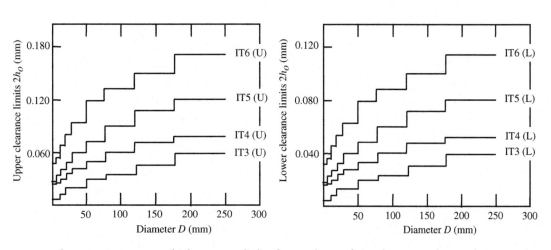

Figure 6.5: Suggested Clearance Limits for Hydrostatic and Aerostatic Bearings.

References

Rowe, W. B., & Stout, K. J. (1975). The design of externally pressurized bearings for reliability with particular reference to manufacturing errors. Proc. 1st European Tribology Congress, London, 1973. *Proceedings of the Institution of Mechanical Engineers, London* 421−429.

Stout, K. J., & Rowe, W. B. (1974). Externally pressurized bearings − Part 3: Design of hydrostatic bearings. *Tribology International, 7*(5; Oct), 195−212.

CHAPTER 7

Plane Hydrostatic and Aerostatic Bearings

Summary of Key Design Formulae

$$A_e = A \cdot \overline{A}$$ Effective pad area

$$W = P_s A_e \overline{W}$$ Load

$$q_o = \frac{P_s h_o^3}{\eta} \cdot \beta \overline{B}$$ Hydrostatic flow

$$q_o = \frac{P_s h_o^3}{\eta} \cdot K_{go} \overline{B} \cdot \left(\frac{p_{ro} + p_a}{2 p_a} \right)$$ Aerostatic flow

$$\lambda = \frac{P_s A_e}{h_o} \cdot \overline{\lambda}$$ Film stiffness

$$h = h_o \cdot \overline{X}$$ Film thickness

7.1 Use of the Design Charts

Chapter 4 gave effective area shape factors and flow shape factors for various plane pads. This chapter gives flow, load, and stiffness data with film thickness varying from the design condition from minimum to maximum. Design charts are for any shape of pad for four situations:

- Single plane pads
- Equal opposed-pad bearings
- Unequal opposed-pad bearings
- Complex configurations.

Gauge pressures are given in upper case symbols. Absolute pressures are given in lower case.

7.2 Choice of Land Width

For most hydrostatic and aerostatic applications it is recommended that the ratio of the land width a to the bearing width L should be $a/L = 0.25$. However, for aerostatic pads, it is desirable to design virtual recesses for reduced recess volume, as described in Section 4.2.

The recommended land-width ratio yields minimum pumping power to support a given load on a given total area.

Hydrostatic, Aerostatic and Hybrid Bearing Design.
DOI: 10.1016/B978-0-12-396994-1.00007-3

At higher speeds, friction power must be taken into account. If required, land-width ratio can be reduced to increase flow for hydrostatic bearings. Reducing land-width ratio may often be preferred to increasing film thickness as a means of increasing flow. It is not recommended to reduce land-width ratio to extreme values. Usually, a land-width ratio of 0.1 is a practicable minimum. Very thin land widths can be vulnerable to damage.

Small land width is not recommended for aerostatic pads with virtual recesses (Section 4.2).

7.3 Flow Variation with Film Thickness

Flow q from a pad varies with film thickness. Hydrostatic flow is given by

$$q = \frac{P_r h^3}{\eta} \cdot \overline{B} = \frac{P_s h_o^3}{\eta} \cdot \overline{B} \cdot \overline{P}_r \cdot \overline{X}^3 \tag{7.1}$$

where $\overline{B}$, the flow shape factor, is the same for hydrostatic and aerostatic bearings. $\overline{P}_r = P_r/P_s$ is given on charts for single pads and $\overline{X} = h/h_o$ takes account of film thickness variation from the design condition. The design condition is defined by $h = h_o$. This yields concentric flow q_o at a recess pressure $P_{ro} = \beta P_s = K_{go} P_s$. The term β is the design pressure ratio for hydrostatic bearings and K_{go} is for aerostatic bearings:

$$q_o = \frac{P_s h_o^3}{\eta} \cdot \beta \overline{B} \qquad\qquad \text{Hydrostatic}$$

$$q_o = \frac{P_s h_o^3}{\eta} \cdot K_{go} \overline{B} \cdot \left(\frac{p_{ro} + p_a}{2 p_a} \right) \qquad \text{Aerostatic} \tag{7.2}$$

Shape factors for a range of pad shapes are provided in Chapter 4.

7.4 Load Variation with Film Thickness

Bearing load $W = P_r A_e$. Effective pad area is $A_e = A \cdot \overline{A}$, where $\overline{A}$ is the area shape factor described in Chapter 4. More generally, for single- and opposed-pad bearings:

$$W = P_s A_e \overline{W} \tag{7.3}$$

where $\overline{W}$ is a load factor that varies with film thickness.

For single pads, load support is $W = P_s A_e \cdot \overline{P}_r$, where $\overline{P}_r = P_r/P_s$. In this case,

$$\overline{W} = \overline{P}_r \qquad \text{Single pads} \tag{7.4a}$$

For opposed pads, $W = P_s A_e \cdot (\overline{P}_{r1} - \overline{P}_{r2})/2$, so that

$$\overline{W} = (\overline{P}_{r1} - \overline{P}_{r2})/2 \qquad \text{Opposed pads} \tag{7.4b}$$

For unequal opposed pads, the load factor takes account of the area ratio of the opposed pads. Charts are given for several methods of flow control.

7.5 Stiffness Variation with Film Thickness

Bearing film stiffness λ is given by $\lambda = (P_s A_e / h_o) \cdot \bar{\lambda}$, where $\bar{\lambda}$ depends on the flow control, as discussed in Chapter 5. Values of $\bar{\lambda}$ are given in this chapter for plane pads as film thickness h varies from the design value h_o. Variations of film thickness are expressed on the charts as $\bar{X} = h/h_o$ varying from 0 to 1. Stiffness varies between a maximum at best to zero at extremes of load. Extreme loads should be avoided. Stiffness factors with capillary control give a better indication of average stiffness within a moderate load range.

7.6 Single-Pad Bearings

Expressions for load, pressure, stiffness, and flow factors are given in Table 7.1 for hydrostatic pads and in Table 7.2 for aerostatic pads. Data charts are presented in Figures 7.1–7.9.

Aerostatic bearings involve more complex expressions than hydrostatic bearings but are still straightforward when employing slot restrictors. Orifice-fed aerostatic bearings are best designed to operate with unchoked orifices to minimize the risk of pneumatic hammer. Figure 7.4b shows the regions where jet flow remains unchoked. The consequence of design for unchoked flow is a reduced range of loads that may be applied to the bearing. Orifice-fed aerostatic pads are better loaded with increasing loads compared with the design condition to avoid choked flow.

Aerostatic pads should be designed to avoid large recess volume, as described in previous chapters. The application of virtual recesses is described in Section 4.2 but virtual recesses reduce load support, requiring slightly increased pad area in some cases. Minimizing recess volume is essential for good dynamic performance.

Table 7.1: Thrust, stiffness, and gap for single hydrostatic pads ($\overline{W} = \overline{P}_r$)

Capillary or Slot	Orifice	Constant Flow	Diaphragm
$\bar{P}_r = \dfrac{1}{1 + \dfrac{1-\beta}{\beta}\bar{X}^3}$	$\bar{X}^6 = \dfrac{1-\bar{P}_r}{\bar{P}^2}\dfrac{\beta^2}{1-\beta}$	$\bar{P}_r = \dfrac{\beta}{\bar{X}^3}$	
$\bar{\lambda} = 3\dfrac{\beta}{1-\beta} \cdot \dfrac{\bar{X}^2}{\left[\bar{X}^3 + \dfrac{\beta}{1-\beta}\right]^2}$		$\bar{\lambda} = \dfrac{3\beta}{\bar{X}^4}$	$\bar{\lambda}_o = \dfrac{3\beta(1-\beta)}{1 - \dfrac{3\beta}{\bar{\lambda}_d}(1-\beta)}$
$\bar{\lambda}\bar{X} = 3\bar{P}_r(1-\bar{P}_r)$	$\bar{\lambda}\bar{X} = 6\bar{P}_r\dfrac{(1-\bar{P}_r)}{2-\bar{P}_r}$	$\bar{\lambda}\bar{X} = 3\bar{P}_r$	
$\bar{Q} = \bar{P}_r \bar{B}\bar{X}^3$	$\bar{Q} = \bar{P}_r\bar{B}\bar{X}^3$	$\bar{Q} = \bar{P}_r\bar{B}\bar{X}^3$	$\bar{Q} = \bar{P}_r\bar{B}\bar{X}^3$

Table 7.2: Thrust, stiffness, and gap for single aerostatic pads

Capillary or Slot	Orifice

$$p_r = \sqrt{\frac{p_s^2 + p_a^2 m_c \overline{X}^3}{1 + m_c \overline{X}^3}}$$

$$\overline{X}^3 = m_o \cdot \frac{\sqrt{(\overline{p}_r^*)^{2/k} - (\overline{p}_r^*)^{(k+1)/k}}}{\overline{p}_r^2 - \overline{p}_a^2}$$

$$\overline{\lambda} = 3 \cdot \frac{m_c \overline{X}^2}{[1 + m_c \overline{X}^3]^2} \cdot \frac{(p_s + p_a)}{2 p_r}$$

$$m_c = \frac{1 - K_{go}}{K_{go}} \cdot \frac{p_s + p_{ro}}{p_{ro} + p_a}$$

$$m_o = \frac{\overline{p}_{ro}^2 - \overline{p}_a^2}{\sqrt{(\overline{p}_{ro}^*)^{2/k} - (\overline{p}_{ro}^*)^{(k+1)/k}}}$$

$$\overline{W} = (p_r - p_a)/(p_s - p_a)$$

$$\overline{p} = \frac{p}{p_s} \quad \text{and} \quad \overline{p}^* = \max\left\{ \overline{p} \text{ or } \left(\frac{2}{k+1}\right)^{k/(k-1)} \right\}$$

$$\overline{Q} = \frac{p_r^2 - p_a^2}{2 p_a (p_s - p_a)} \overline{B} \overline{X}^3$$

$$\overline{Q} = \frac{p_r^2 - p_a^2}{2 p_a (p_s - p_a)} \overline{B} \overline{X}^3$$

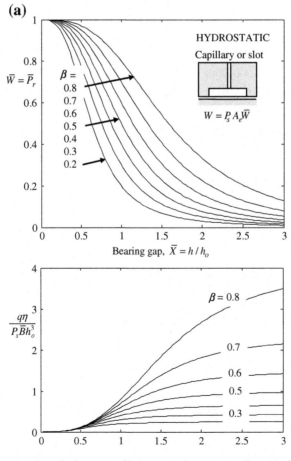

(a)

Figure 7.1: (a) Load and Flow: Capillary and Slot-Controlled Hydrostatic Pads.

(b)

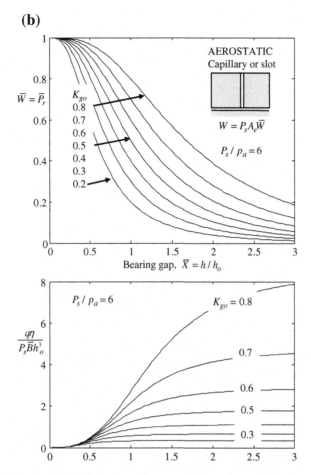

Figure 7.1: (b) Load and Flow: Capillary and Slot-Controlled Aerostatic Pads.

Selection of Design Pressure Ratio $\beta = K_{go} = P_{ro}/P_s$ for Single Pads

The general recommendation is $\beta = K_{go} = 0.5$. This allows the maximum range of applied loads. Higher or lower values may be employed for special cases. However, for orifice-fed aerostatic bearings, pressure ratio should not be less than 0.5 to avoid choked flow and possible instability.

Adjusting Flow Rate to a Convenient Value

Often, flow for a low-speed single circular pad is very small. Or, sometimes, flow is too high for the pump proposed. There are a number of ways to reduce flow, such as designing thicker lands or increasing bearing size while reducing supply pressure. Alternatively, if bearing area may be increased it may be possible to reduce pressure ratio, β, to about 0.2, which has the advantage that a higher maximum load may be applied. However, the most sensitive parameter controlling flow is gap h because it appears to the third power.

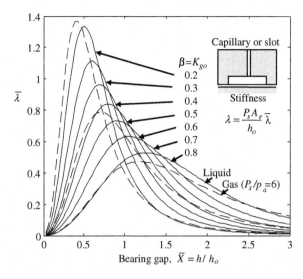

Figure 7.2: Stiffness and Gap for Capillary and Slot-Controlled Pads.

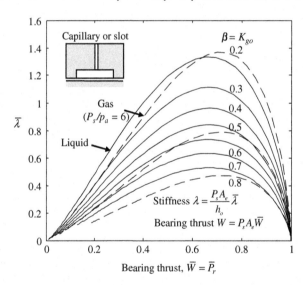

Figure 7.3: Stiffness and Thrust for Capillary-Controlled Pads.

■ *Example 7.1 A Single-Pad Hydrostatic Bearing*

A circular capillary-compensated pad bearing with central oil admission must support a load of 4.5 kN (1011 lbf). The bearing gap is to be 50 μm (0.002 in) and the bearing is to operate at zero speed. Determine:

1. Suitable dimensions for the pad
2. Operating stiffness

(a)

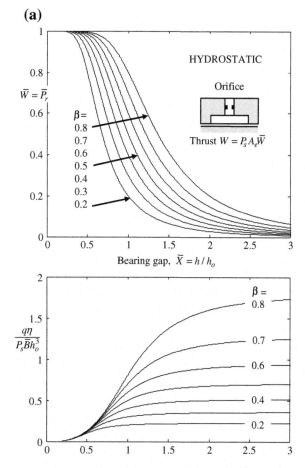

Figure 7.4: (a) Load and Flow for Orifice-Controlled Hydrostatic Pads.

3. Load support and flow when $\overline{X}$ is reduced to 0.4
4. Load when $h = 10$ μm (0.0004 in).

It has been decided that β is to be 0.5 and a convenient pressure for the pump is 4 MN/m^2 (580 lbf/in^2). The oil to be used has a dynamic viscosity $\eta = 35$ cP (5×10^{-6} reyn) at 38 °C (100 °F).

Solution

1. At the design condition $P_r = 0.5P_s = 0.5 \times 4 = 2$ MN/m^2 (290 lbf/in^2)

$$A_e = A \cdot \overline{A} = \frac{W}{P_r} = 4500/(2 \times 10^6) = 2.25 \times 10^{-3} \text{ m}^2 \text{ (3.487 in}^2)$$

(b)

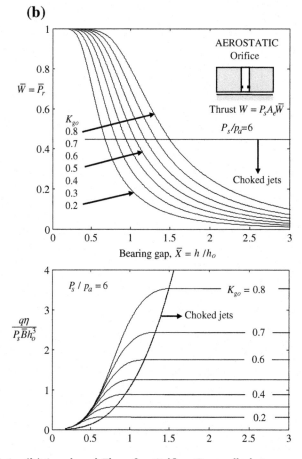

Figure 7.4: (b) Load and Flow for Orifice-Controlled Aerostatic Pads.

From Figure 4.2 and choosing $a/D = 0.25$, so that $\overline{R} = R_2/R_1 = 2.0$ and $\overline{A} = 0.542$,

$$A = \frac{A\overline{A}}{\overline{A}} = \pi R_2^2 = \frac{2.25 \times 10^{-3}}{0.542} = 4.15 \times 10^{-3} \text{ m}^2 \ (6.435 \text{ in}^2)$$

$R_2 = 36.3$ mm (say 36 mm) (1.42 in)

$R_1 = R_2/\overline{R} = 36/2 = 18$ mm (0.71 in)

The design gap is achieved by setting the flow. From Figure 4.2, $\overline{B} = 0.75$, therefore

$$q_o = \frac{P_s h_o^3}{\eta} \beta \overline{B} = \frac{4 \times 10^6 \times 50^3 \times 10^{-18} \times 0.5 \times 0.75}{0.035}$$

$$= 5.36 \times 10^{-6} \text{ m}^3/\text{s} \ (0.327 \text{ in}^3/\text{s})$$

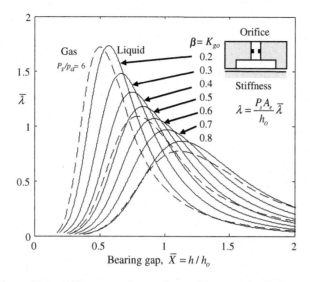

Figure 7.5: Stiffness and Gap for Orifice-Controlled Pads.

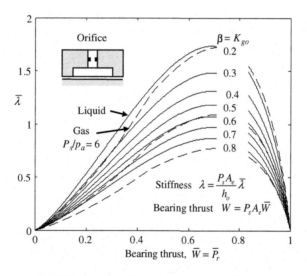

Figure 7.6: Stiffness and Thrust for Orifice-Controlled Pads.

2. From Figure 7.2, for $\beta = 0.5$ and $\overline{X} = 1$, the stiffness factor $\overline{\lambda} = 0.75$. Actual stiffness is therefore

$$\lambda = \frac{P_s A \overline{A}}{h_o} \overline{\lambda} = \frac{4 \times 10^6 \times 2.25 \times 10^{-3} \times 0.75}{50 \times 10^{-6}} = 0.135 \text{ GN/m } (0.771 \times 10^6 \text{ lbf/in})$$

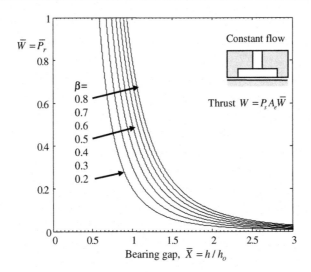

Figure 7.7: Load and Gap for Constant-Flow Controlled Pads.

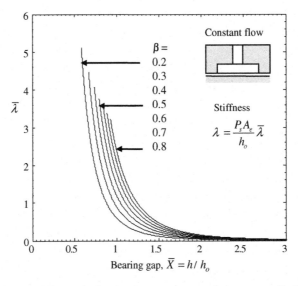

Figure 7.8: Stiffness and Gap for Constant-Flow Controlled Pads.

3. When $\overline{X} = 0.4$, $h = 50 \times 0.4 = 20 \ \mu$m (0.0008 in).
    From Figure 7.1, the load factor $\overline{W} = 0.94$. Therefore, maximum applied load is

    $$W = P_s A \overline{A} \ \overline{W} = 4 \times 10^6 \times 2.25 \times 10^{-3} \times 0.94 = 8.46 \text{ kN (1900 lbf)}$$

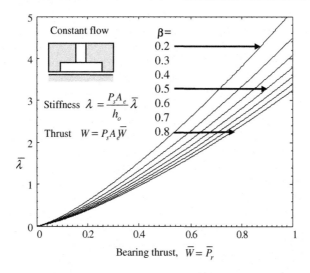

Figure 7.9: Stiffness and Thrust for Constant-Flow Controlled Pads.

Maximum load at this gap compared with the design load is $8460/4500 = 1.88$ and the dimensionless stiffness is reduced to 0.42 compared with the design value of 0.75.

From Figure 7.1 or Table 7.1, dimensionless flow is $\overline{Q}_o = \beta \overline{B} = 0.5 \times 0.75 = 0.375$ at the design gap. At the reduced film thickness, dimensionless flow is reduced to $\overline{Q} = \overline{B}\,\overline{P}_r \overline{X}^3 = 0.75 \times 0.94 \times 0.064 = 0.0451$ and actual flow is

$$q = \frac{P_s h_o^3}{\eta} \cdot 0.0451 = \frac{4 \times 10^6 \times 50^3 \times 10^{-18} \times 0.0451}{0.035}$$

$$= 0.6443 \times 10^{-6} \text{ m}^3/\text{s} \ (0.0393 \text{ in}^3/\text{s})$$

4. Taking $h = 10 \ \mu$m (0.0004 in) as the minimum allowable film thickness,

$$\overline{X} = h/h_o = 10/50 = 0.2 \quad \text{and} \quad \overline{W} = 0.99$$
$$W = P_s A \overline{A} \cdot \overline{W} = 4 \times 10^6 \times 2.25 \times 10^{-3} \times 0.99 = 8.91 \text{ kN (2000 lbf)}$$

However, dimensionless stiffness is now reduced to 0.1. This demonstrates that minimum allowable film thickness is not a good criterion for maximum allowable load. A better approach is to decide minimum $\overline{X}$. For $\beta = 0.5$ a suitable range for $\overline{X}$ is from 0.4 to 1.6. If, however, $\beta = 0.2$, a suitable range for $\overline{X}$ is from 0.2 to 1.1.

∎

■ *Example 7.2 A Single-Pad Aerostatic Bearing*

An aerostatic slideway bearing must support a load of 2 kN (450 lbf). The bearing gap is to be 25 μm (0.001 in).

It has been decided that K_{go} is to be 0.5 and a convenient air pressure is 0.45 MN/m^2 (65.3 lbf/in^2). Air has a dynamic viscosity $\eta = 18.3 \times 10^{-6}$ Ns/m^2 at 18 °C. Ambient pressure is 0.101 MN/m^2 (14.7 lbf/in^2).

Determine:

1. Suitable dimensions for the pad and entry ports
2. Required flow at the design condition
3. Load support when the film thickness is reduced by 50%
4. Stiffness when the film thickness is reduced by 50% employing capillary flow control.

It is decided to choose a rectangular pad with a virtual recess. This requires a number of holes or entry ports to be arranged around the virtual recess.

Solution

1. To determine suitable dimensions. Increase the specified load by 15% using virtual recesses to allow for dispersion from discrete entry ports: $W_o = 2300$ N.

$$\text{See Figure 4.4 for shape factors}: B/L = 2, \; a/L = 0.3, \; \overline{A} = 0.57, \overline{B} = 1.2$$

$$\text{At the design condition,} \; P_{ro} = 0.5 P_s = 0.5 \times 0.45 \times 10^6$$
$$= 0.225 \; \text{MN/m}^2 \; (32.7 \; \text{lbf/in}^2).$$

$$W = P_r A \cdot \overline{A} \text{ so that } A\overline{A} = 2300/(0.225 \times 10^6) = 10.22 \times 10^{-3} \; \text{m}^2 \; (15.84 \; \text{in}^2)$$

$$A = 10.22 \times 10^{-3}/0.57 = 0.01793 \; \text{m}^2 \; (27.79 \; \text{in}^2)$$

$$\frac{B}{L} \cdot L = 2L^2 = 0.01793 \; \text{m}^2$$

$$L = \frac{(\sqrt{0.01793})}{2} = 0.06695 \; \text{m} \; (2.636 \; \text{in})$$

$$B = 2L = 0.1339 \; \text{m} \; (5.272 \; \text{in})$$

$$a = 0.3 \times 0.06695 = 0.0201 \; \text{m} \; (0.791 \; \text{in})$$

Assume the virtual recess will be surrounded at least 70% by slot-entry ports. The perimeter of the virtual recess is

$$2(B + L - 4a) = 2(0.1339 + 0.06695 - 4 \times 0.0201) = 0.2409 \text{ m } (9.484 \text{ in})$$

For 70% recess surround:
Total length of the entry slots must be $0.7 \times 0.2409 = 0.169$ m (6.64 in).
Assume 16 slot shaped entry ports.
The length of the slots must be 0.0105 m (0.415 in).
Suggested port dimensions are 1 mm wide $\times$ 0.1 mm deep.
Total port volume is $16 \times 0.011 \times 0.001 \times 0.0001 = 17.6 \times 10^{-9}$ m^3. The suitability of this port volume can be compared with the flow. See next step.

2. To determine free air flow at the design condition.
Absolute recess pressure is $p_r = 0.225 + 0.101 = 0.326$ MN/m^2 (47.3 lbf/in^2). $\overline{B} = 1.2$.
From equation (7.2):

$$q_o = \frac{P_s h_o^3}{\eta} K_{go}\overline{B}\left(\frac{p_{ro} + p_a}{2p_a}\right)$$

$$q_o = \frac{0.45 \times 10^6 \times 25^3 \times 10^{-18} \times 0.5 \times 1.2}{0.0000183}\left(\frac{0.326 + 0.101}{2 \times 0.101}\right)$$

$$= 487 \times 10^{-6} \text{ m}^3/\text{s } (0.487 \text{ l/s})$$

This is the flow for slot-entry ports that 100% surround the virtual recess. Flow may be reduced by approximately 15% employing a virtual recess with 70% surround. The ratio of the flow to the port volume is $f = (487 \times 10^{-6})/(17.6 \times 10^{-9}) = 27.7 \times 10^3$ Hz. See Equation (5.16). This frequency is much larger than 1000 Hz, which suggests the size of the entry ports should not cause a problem, although this is not a guarantee.

3. To determine load when film thickness is reduced by 50%.

$$\overline{X} = \frac{h}{h_o} = 0.5$$

From Figure 7.1, $\overline{W} = 0.88$. Maximum applied load for this gap is

$$W = P_s A \overline{A}\,\overline{W} = 0.45 \times 10^6 \times 0.01793 \times 0.57 \times 0.88 = 4.05 \text{ kN } (910 \text{ lbf})$$

The shape of the graph suggests that this load represents an absolute maximum that should be applied under any circumstances.

4.   From Figure 7.2, for $\beta = 0.5$ and $\overline{X} = 0.5$, dimensionless stiffness $\overline{\lambda} = 0.62$. Stiffness is

$$\lambda = \frac{P_s A \overline{A}}{h_o} \; \overline{\lambda} = \frac{0.45 \times 10^6 \times 0.01793 \times 0.57 \times 0.62}{25 \times 10^{-6}} = 114 \, \text{MN/m} \; (0.65 \times 10^6 \, \text{lbf/in})$$

■

7.7 Equal Opposed-Pad Bearings

The design of hydrostatic bearing systems having equal opposed pads is simpler in practice than for single-pad bearings. Data are given below in Figures 7.10–7.17. For aerostatic

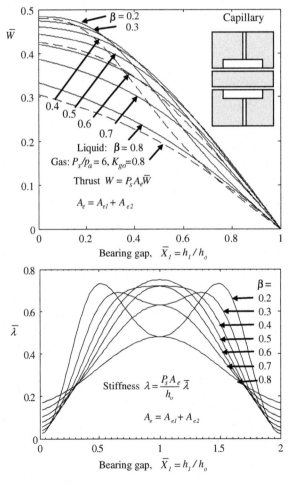

Figure 7.10: Load and Stiffness for Capillary-Controlled Equal Opposed Pads.

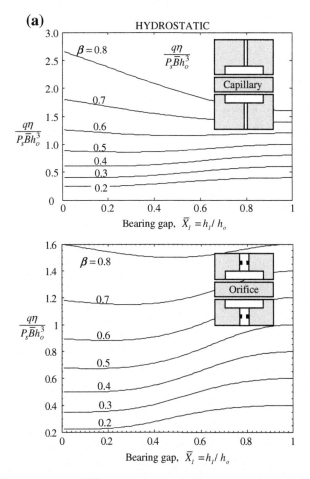

Figure 7.11: (a) Flow: Hydrostatic Equal Opposed Pads.

bearings, the same charts can be employed as for hydrostatic bearings. The differences are relatively minor. Dashed lines are included for aerostatic bearings to show where reduced load support applies compared with hydrostatic bearings. The reduction is greatest for low design pressure ratio.

The effective area of an equal opposed-pad bearing is twice the effective area of each of the single pads employed to make up the bearing. In other words, $A\overline{A} = A_e = A_{e1} + A_{e2}$, where 1 and 2 denote the opposing pads. Load W, stiffness λ, and flow q_o are given by

$$W = P_s A_e \overline{W} \quad \text{where} \quad A_e = A_{e1} + A_{e2} \tag{7.5}$$

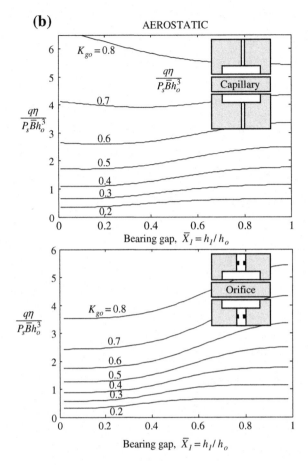

Figure 7.11: (b) Flow: Aerostatic Equal Opposed Pads, $P_s/p_a = 6$.

$$\lambda = \frac{P_s A_e}{h_o}\overline{\lambda} \quad \text{where} \ A_e = A_{e1} + A_{e2} \tag{7.6}$$

$$q_o = \frac{P_s h_o^3}{\eta} \cdot 2\beta\overline{B} \qquad\qquad \text{Hydrostatic}$$

$$q_o = \frac{P_s h_o^3}{\eta} \cdot 2K_{go}\overline{B}\left(\frac{p_r + p_a}{2p_a}\right) \qquad \text{Aerostatic} \tag{7.7}$$

Values are given on the data charts that follow where film thicknesses of the two pads are

$$\overline{X}_1 = h_1/h_o \quad \text{and} \quad \overline{X}_2 = 2 - \overline{X}_1 \tag{7.8}$$

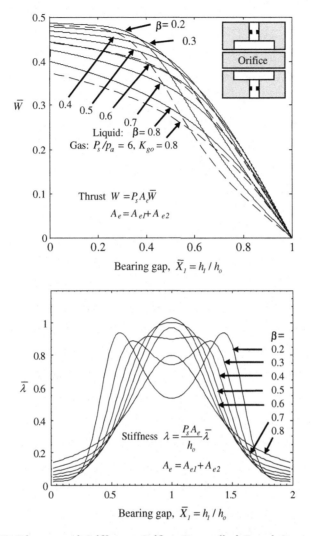

Figure 7.12: Thrust and Stiffness: Orifice-Controlled Equal Opposed Pads.

Selection of Design Pressure Ratio for Opposed-Pad Bearings

The optimum design pressure ratio is $\beta = K_{go} = 0.5$ and in most cases it is unnecessary to choose any other value for opposed-pad bearings.

Where tolerances are given on clearance, a suitable range for pressure ratio is $0.4 < \beta < 0.7$ or for orifice-fed aerostatic bearings a suitable reduced range is $0.5 < K_{go} < 0.7$. As explained in previous chapters, it is generally recommended that choked flow is avoided for orifice-fed aerostatic bearings to reduce the risk of instability.

With pressure-sensing flow control devices the maximum allowable load will not be lower than that for capillary compensation and therefore the same value may be used.

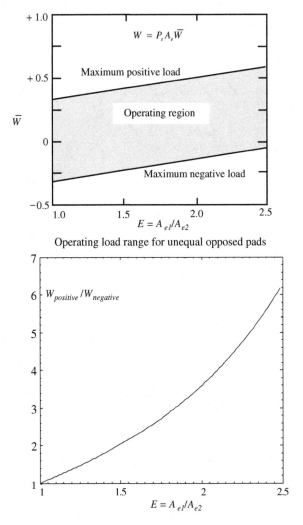

Figure 7.13: Unequal Opposed Pads: Operating Load Range and Selection of Area Ratio.

■ *Example 7.3 An Opposed-Pad Hydrostatic Bearing*

A capillary-compensated opposed-pad bearing is required that will not deflect more than 2.5 μm (0.0001 in) under an applied load of 450 N (101 lbf). The available supply pressure is 3.5 MN/m² (508 lbf/in²) and in the design condition there is negligible load. The design bearing gap is to be $h_{o1} = h_{o2} = 25$ μm (0.001 in). The pressure ratio β is to be 0.5. Determine:

1. Suitable bearing dimensions for circular bearing pads
2. The maximum allowable applied load.

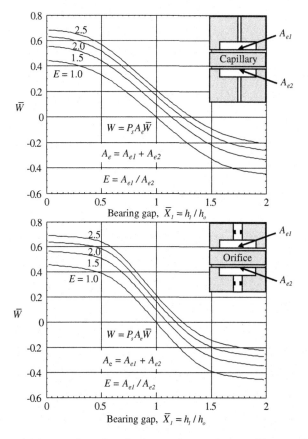

Figure 7.14: Unequal Opposed Pads: Thrust and Gap for Capillary and Orifice Control.

Solution

1. *Bearing dimensions.* Stiffness must be at least $450/2.5 \times 10^{-6} = 180$ MN/m (1.03×10^6 lbf/in). From Figure 7.10, stiffness factor is $\bar{\lambda} = 0.75$. Actual stiffness is given by

$$\lambda = 180 \times 10^6 = \frac{P_s A_e}{h_o}\bar{\lambda} = \frac{3.5 \times 10^6 \times A_e}{25 \times 10^{-6}} \times 0.75$$

A_e must be greater than 1714×10^{-6} m^2, $A_{e1} = A_{e2}$ must be greater than 857×10^{-6} m^2. From Figure 4.2, read $\bar{A} = 0.54$ at $R_2/R_1 = 2$:

$$A_1 = \frac{A_{e1}}{\bar{A}} = \frac{857 \times 10^{-6}}{0.54} = 1587 \times 10^{-6} \text{ m}^2 = \pi R_2^2$$

$$R_2 = 22.5 \text{ mm (0.886 in)}$$

$$R_1 = 11.25 \text{ mm (0.443 in)}$$

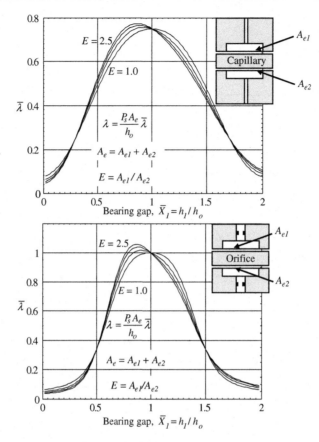

Figure 7.15: Unequal Opposed Pads: Stiffness and Gap for Capillary and Orifice Control.

2. *Maximum allowable load.* Examining Figure 7.10, it is seen that stiffness is greatly reduced for $\overline{W}$ greater than 0.3. A sensible maximum load would be

$$\overline{W} = P_s A_e \overline{W} = 3.5 \times 10^6 \times 1714 \times 10^{-6} \times 0.3 = 1.8 \text{ kN (407 lbf)}$$

■

■ *Example 7.4 An Opposed-Pad Aerostatic Bearing*

An aerostatic capillary-compensated opposed-pad bearing is to have a design bearing film thickness $h_{o1} = h_{o2} = 25 \ \mu$m (0.001 in). The maximum applied load is to be 450 N (101 lbf). It is required that the gap will not reduce by more than 50%. The available supply pressure is 0.45 MN/m^2 (65 lbf/in^2) and in the design condition there is negligible load. The pressure ratio is to be $K_{go} = 0.5$. Determine suitable dimensions for circular bearing pads and check flow rate.

Solution

1. *Bearing dimensions.* Fifty percent minimum film thickness corresponds to $\bar{X} = 0.5$. From Figure 7.10, load factor $\overline{W} \approx 0.3$. Increasing required load by 15% to allow for losses due to discrete entry ports:

$$W = 517.5 \text{ N } (116.3 \text{ lbf})$$

Effective bearing area is $A\overline{A} = \dfrac{W}{P_s \overline{W}} = \dfrac{517.5}{0.45 \times 10^6 \times 0.3} = 0.003833 \text{ m}^2 \ (5.941 \text{ in}^2)$

Assume opposed circular pads with a virtual recess and $a/D = 0.3$, so that $R_2/R_1 = 2.5$. From Figure 4.2, $\overline{A} = 0.54$ and $\overline{B} = 0.57$,

$A = 0.003833/0.54 = 0.007098 \text{ m}^2 \ (11 \text{ in}^2)$ total area of two opposing pads

$A_1 = A_2 = 0.5A = 0.0003549 \text{ m}^2 \ (5.5 \text{ in}^2)$ area of each pad

$D = \sqrt{\dfrac{4 \times 0.003549}{\pi}} = 0.06722 \text{ m } (2.647 \text{ in}^2)$ pad outer diameter

$D_r = 0.06722 \times 0.4 = 0.02689 \text{ m } (1.059 \text{ in})$ virtual recess diameter

Perimeter of virtual recess is $\pi \times 0.002689 = 0.08448 \text{ m } (3.326 \text{ in})$

Assume eight small-entry recesses. For 70% surround of the large virtual recess, the entry recess diameter or groove length:

$$d_r = 0.7 \times 0.08448/8 = 0.007392 \text{ m } (0.291 \text{ in})$$
$$h_r = 0.5 \text{ mm } (0.0197 \text{ in}) \text{ depth of recess}$$

Recess volume for two pads is

$V_r = 2 \times 8 \times \pi \times 0.007392^2 \times 0.0005/4 = 0.3434 \times 10^{-6} \text{ m}^3 \ (0.02096 \text{ in}^3)$

2. *Flow.* From Chapter 4, the flow shape factor for a circular pad where $D/D_r = 2.5$ is $\overline{B} = 0.57$:

Flow $q_o = 2\dfrac{P_s h_o^3}{\eta} K_{go}\overline{B} \dfrac{p_{ro} + p_a}{2p_a}$ for two opposed pads

$p_{ro} = 0.225 + 0.101 = 0.326 \text{ MN/m}^2; \quad p_a = 0.101 \text{ MN/m}^2$

$q_o = 2 \times \dfrac{0.45 \times 10^6 \times 25^3 \times 10^{-18}}{18.3 \times 10^{-6}} \times 0.5 \times 0.57 \times \dfrac{0.326 + 0.101}{2 \times 0.101}$

$= 463 \times 10^{-6} \text{ m}^3/\text{s } (0.463 \text{ l/s})$

Flow may be reduced approximately 15% employing a virtual recess with 70% surround. Check recess volume from equation (5.16):

$$f = \frac{q_o}{V_r} = \frac{463 \times 10^{-6}}{0.3434 \times 10^{-6}} = 1348 \text{ Hz}.$$

At first sight, this looks sufficiently high.

■

7.8 Unequal Opposed-Pad Bearings

Unequal opposed-pad bearings may be used to advantage where the principal forces are in one direction but where forces in the reverse direction may also occur. At speed there is the additional advantage of lower power losses and temperature rise than with the equivalent equal opposed-pad configuration.

The basic schematic arrangement is apparent from the design charts (Figures 7.14 and 7.15), which are intended to be self-explanatory. One side of the bearing, usually the top, carries the main load and is made bigger than the other side. Thus, if the top bearing has an effective area A_{e1} and the lower has an effective area A_{e2}, the bearing has an area ratio $E = A_{e1}/A_{e2}$.

The result of the difference between the top and bottom areas is that the bearing carries a load at the design operating gap where $\overline{X} = 1$. Another consequence is that the maximum positive bearing thrust $W_{positive}$ in the upward direction is larger than the maximum negative bearing thrust $W_{negative}$ in the downward direction.

The bearing must therefore be designed so that the range of applied loads lie within the acceptable operating range. The charts in Figure 7.13 provide an approximate guide to an acceptable load range.

Charts are given for bearing thrust with gap variation for both capillary and orifice control in Figure 7.14. The area ratios cover a range from $E = 1$ to 2.5, which should be adequate for most purposes.

The charts are all computed for a value of the design pressure ratio $\beta = P_{ro}/P_s = 0.5$.

The load and stiffness relationships are the same as previous cases for equal opposed pads:

$$W = P_s A_e \overline{W} \quad \text{where } A_e = A_{e1} + A_{e2} \tag{7.9}$$

and

$$\lambda = \frac{P_s A_e}{h_o} \, \overline{\lambda} \quad \text{where} \quad A_e = A_{e1} + A_{e2} \tag{7.10}$$

Charts of stiffness data for capillary and orifice control are given in Figure 7.15.

Flow at the design condition is the sum of the flows for the two pads:

$$q_o = \frac{P_s h_o^3}{\eta} (\beta \overline{B}_1 + \beta \overline{B}_2) \tag{7.11}$$

It is assumed that the proportions of the two pads will be the same, only the size will be different, so that pad shape factors $\overline{A}$ and $\overline{B}$ are identical for the two pads. Flow data may therefore be read from Figure 7.11, which gives flow charts for equal opposed pads. Pad size and area ratio have no effect on flow data as long as the pad shapes are identical on each side of the bearing.

The shape of the characteristic curves for unequal opposed pads depends on the ratio of the areas of the two opposing pads and is termed the area ratio E:

$$E = \frac{A_{e1}}{A_{e2}} \tag{7.12}$$

A design procedure is given below, which is intended to be self-explanatory as far as possible, and an example is given to illustrate how the procedure can be used.

The operating load range of unequal opposed-pad bearings is indicated in Figure 7.13 for various area ratios. It was assumed that it would be undesirable to operate with stiffness lower than approximately half the stiffness value at the design condition. This condition ensures a small margin for overload before bearing collapse occurs. The information in Figure 7.13 is presented in a form convenient for use in the design procedure. Area ratio E is indicated for a given ratio of maximum positive load to maximum negative load.

The charts are all computed for design pressure ratio $\beta = 0.5$. Other values of design pressure ratio are not considered to offer significant advantage.

7.9 Complex Arrangements of Pads (Capillary Controlled)

Complex bearing arrangements can be analyzed without too much difficulty for a four-pad problem such as the example in Figure 7.16. Analysis can be carried out for even more pads if there is symmetry or if some pads are arranged perpendicular to others.

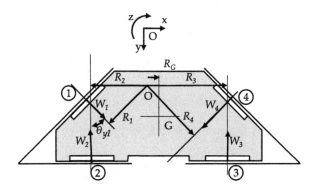

Figure 7.16: Bearing Coordinates for a Complex Arrangement of Pads.

Given a set of bearing deflections or gaps, it is a simple matter to find the bearing thrusts. In practical problems, it is more often the other way round. One has to find the deflections from a known set of applied loads. Unfortunately, there is no simple method.

The designer usually overcomes this problem by using the following approximate analysis for deflections and ensuring a margin of safety.

The procedure is for capillary compensation and will ensure a safe design even for other types of compensation, since load capacity will not be greatly altered and operating stiffness will be improved.

Procedure

The procedure basically involves two stages:

1. Calculate pad thrusts at the design condition to ensure a suitable design pressure ratio β and effective pad area A_e for each pad.
2. Calculate approximate dimensionless deflections for the maximum operating loads using the design stiffness values. The dimensionless deflections should represent gap changes that are safe. For example, dimensionless gap should lie within an approximate range between 0.5 and 1.5. If the dimensionless deflections are excessive, the alternatives are to choose a new A_e and a new β or to rearrange the bearing layout and repeat steps 1 and 2.

In Figure 7.16, the pad thrusts at the design condition are shown as W_1, W_2, W_3, and W_4, where

$$W_1 = P_s A_{e1} \beta_1$$
$$W_2 = P_s A_{e2} \beta_2, \text{etc.}$$

We assume that loads F_x, F_y, and F_z applied on the bearing are resolved into three coordinates centered at some point O and act in a two-dimensional plane. For example, a gravity load G acting at a radius R_G creates a moment $F_z = G \times R_G$. Of course, the applied loads will usually act in three dimensions and involve six coordinates. The analysis can be extended to the more complex situation using the same notation.

Applied loads must balance pad thrusts, so that

$$F_x = W_1\cos\theta_{x1} + W_2\cos\theta_{x2} + W_3\cos\theta_{x3} + W_4\cos\theta_{x4}$$
$$F_y = W_1\cos\theta_{y1} + W_2\cos\theta_{y2} + W_3\cos\theta_{y3} + W_4\cos\theta_{y4} \qquad (7.13)$$
$$F_z = W_1R_1 + W_2R_2 + W_3R_3 + W_4R_4$$

Equation (7.13) represents three equations with four unknown A_e values and four unknown β values. It is therefore necessary to choose or guess five of these values arbitrarily. For example, choosing A_{e1}, A_{e2}, A_{e3}, A_{e4} and one β value, it is then possible to solve for the remaining β values. If the β values do not lie in the range from 0.2 to 0.75, some rearrangement of the areas is required.

To analyze deflections:

$$\text{deflection} = \frac{\text{change in load}}{\text{average stiffness}}$$

$$\qquad (7.14)$$

$$h_o - h = \frac{W - W_o}{\overline{\lambda}_o} \quad \text{for small deflections}$$

This may be stated in dimensionless form as

$$1 - \overline{X} = \frac{\overline{W} - \overline{W}_o}{\overline{\lambda}_o} \qquad (7.15)$$

so that

$$W - W_o = P_sA_e\overline{\lambda}_o(1 - \overline{X}) \qquad (7.16)$$

Equation (7.13) can be rewritten for maximum loading to yield approximate dimensionless deflections (a prime symbolizes maximum loading):

$$F'_x - F_x = (W'_1 - W_1)\cos\theta_{x1} + (W'_2 - W_2)\cos\theta_{x2} + (W'_3 - W_3)\cos\theta_{x3}$$
$$+ (W'_4 - W_4)\cos\theta_{x4}$$
$$F'_y - F_y = (W'_1 - W_1)\cos\theta_{y1} + (W'_2 - W_2)\cos\theta_{y2} + (W'_3 - W_3)\cos\theta_{y3} \qquad (7.17)$$
$$+ (W'_4 - W_4)\cos\theta_{y4}$$
$$F'_z - F_z = (W'_1 - W_1)R_1 + (W'_2 - W_2)R_2 + (W'_3 - W_3)R_3 + (W'_4 - W_4)R_4$$

Substituting in equations (7.17) from equation (7.16) yields three equations in four unknown values of $(1 - \overline{X})$. Reducing the number of unknowns is often facilitated by symmetry in the bearing layout, as in the following design example. However, for the general case, the number of unknowns is reduced to three by solving in terms of the three coordinate movements:

$$1 - \overline{X}_1 = \frac{x}{h_{o1}\cos\theta_{x1}} + \frac{y}{h_{o1}\cos\theta_{y1}} + \frac{zR_1}{h_{o1}}$$

$$1 - \overline{X}_2 = \frac{x}{h_{o2}\cos\theta_{x2}} + \frac{y}{h_{o2}\cos\theta_{y2}} + \frac{zR_2}{h_{o2}}$$

(7.18)

and so on.

It is preferable that each bearing should preload another, so that a larger bearing area can be employed and deflections will be small. In Example 7.5, bearing pad 1 is not preloaded and the load range is limited. In many cases, for economic or other reasons, it is not possible to employ pads in opposition. Under such circumstances, the following rules tend to give confidence.

Choice of Design Pressure Ratio P_{ro}/P_s for Complex Pad Arrangements

The optimum is $\beta = K_{go} = 0.5$ but where the greatest load-bearing capacity is required without serious loss of stiffness, the following values are chosen:

1. For bearings where the range of applied loads is small, choose $\beta = 0.5$.
2. For heavily loaded pads that operate at clearances *less* than the design value, choose β in the range 0.2–0.4.
3. For lightly loaded pads that operate at clearances *greater* than the design value, choose β in the range 0.7–0.75.

■ Example 7.5 A Complex Hydrostatic Pad Arrangement

Choose suitable values of the effective area and the pressure ratio for each pad. For the complex pad arrangement shown in Figure 7.17, the supply pressure is 3.5 MN/m² (508 lbf/in²). The loading conditions are:

1. Light loading condition:

$$F_y = 5.5 \text{ kN } (1012 \text{ lbf})$$
$$F_z = 340 \text{ Nm } (3009 \text{ lbf in})$$

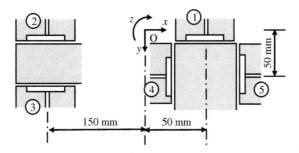

Figure 7.17: Design Example: A Complex Pad Arrangement.

2. Extreme loading condition:

$$F'_x = 5 \text{ kN } (1124 \text{ } lbf)$$
$$F'_y = 12 \text{ kN } (2698 \text{ lbf})$$
$$F'_z = 900 \text{ Nm } (7966 \text{ lbf in})$$

Solution

Select initial pad areas that appear reasonable in the light of the extreme loading condition. In this case pad 1 needs to be large enough to support half the y-loading and a substantial proportion of the z-loading. Initial consideration suggests the following pad areas:

$$A_{e1} = 8000 \text{ mm}^2 \text{ } (12.4 \text{ in}^2)$$
$$A_{e2} = A_{e3} = 2A_{e1} = 16,000 \text{ mm}^2 \text{ } (18.6 \text{ in}^2)$$
$$A_{e4} = A_{e5} = A_{e1} = 8000 \text{ mm}^2 \text{ } (12.4 \text{ in}^2)$$

Using the initial selection of areas, the light and extreme loading situations must now be analyzed and solved to find appropriate values of design pressure ratio. It is possible that the pad areas may need to be adjusted and the procedure repeated until a suitable combination is found.

Analyzing for light loads as a design condition
x direction:

$$F_x = W_4 - W_5$$
$$0 = 3.5 \times 10^6 \times 8 \times 10^{-3}(\beta_4 - \beta_5)$$
$$\beta_4 = \beta_5 \text{ and choose a value of } 0.5$$

y direction:

$$F_y = W_1 \cos y_1 + W_2 \cos y_2 + W_3 \cos y_3 = W_1 + W_2 - W_3$$

where $W_n = P_s A_{en} \beta_n$. Inserting values for the applied load F_y and the pad thrusts W_1, W_2, and W_3 yields

$$5.5 \times 10^3 = 3.5 \times 10^6 \times 8 \times 10^{-3}(\beta_1 + 2\beta_2 - 2\beta_3)$$
$$\beta_1 + 2\beta_2 - 2\beta_3 = 0.1964$$

z direction:

$$F_z = 0.05W_1 - 0.15W_2 + 0.15W_3 - 0.05W_4 + 0.05W_5$$

But $W_4 = W_5$ since $F_x = 0$, therefore

$$340 = 3.5 \times 10^6 \times 8 \times 10^{-3}(0.05\beta_1 - 2 \times 0.15 \times \beta_2 + 2 \times 0.15 \times \beta_3)$$
$$\beta_1 - 6\beta_2 + 6\beta_3 = 0.2429$$

Since there are now two equations in three unknown β values, one value must be chosen arbitrarily, say $\beta_2 = 0.5$. Solving for the x and y directions simultaneously:

$$\beta_3 = 0.506, \text{ say } 0.51$$
$$\beta_1 = 0.208, \text{ say } 0.21$$

Analysis of approximate deflections from equations (7.16) and (7.17) and Figure 7.2
x direction:

$$5 \times 10^3 = 3.5 \times 10^6 \times 8 \times 10^{-3}[0.75(1 - \overline{X}_4) - 0.75(1 - \overline{X}_5)]$$
$$(1 - \overline{X}_5) = -(1 - \overline{X}_4)$$
$$1 - \overline{X}_4 = 0.119$$

y direction:

$$6.5 \times 10^3$$
$$= 3.5 \times 10^6 \times 8 \times 10^{-3}[0.5(1 - \overline{X}_1) + 2 \times 0.75(1 - \overline{X}_2) - 2 \times 0.75(1 - \overline{X}_3)]$$

Since $(1 - \overline{X}_3) = -(1 - \overline{X}_2)$, the equation simplifies to

$$0.232 = 0.5(1 - \overline{X}_1) + 3(1 - \overline{X}_2)$$

z direction:

$$560 = 0.05(W_1' - W_1) - 0.15(W_2' - W_2) + 0.15(W_3' - W_3)$$
$$-0.05(W_4' - W_4) + 0.05(W_5' - W_5)$$
$$560 = 3.5 \times 10^6 \times 8 \times 10^{-3}[0.05 \times 0.5(1 - \overline{X}_1) - 2 \times 0.15 \times 0.75(1 - \overline{X}_2)$$
$$+2 \times 0.15 \times 0.75(1 - \overline{X}_3) - 0.05 \times 0.75(1 - \overline{X}_4) + 0.05 \times 0.75(1 - \overline{X}_5)]$$
$$0.4 = 0.5(1 - \overline{X}_1) - 6 \times 0.75(1 - \overline{X}_2) + 6 \times 0.75(1 - \overline{X}_3)$$
$$-0.75(1 - \overline{X}_4) + 0.75(1 - \overline{X}_5)$$
$$1 - \overline{X}_4 = 0.119 \text{ from above}$$
$$0.5785 = 0.5(1 - \overline{X}_1) - 9(1 - \overline{X}_2)$$

Solving simultaneously from the y and z directions yields the summary of bearing gap values at the extreme load condition:

$$(1 - \overline{X}_2) = -0.0289$$
$$(1 - \overline{X}_1) = 0.2907$$

Summarizing the resulting values of β and approximate film thickness $\overline{X}$ at extreme loading:

$$
\begin{aligned}
\beta_1 &= 0.21 & \overline{X}_1 &= 0.709 \\
\beta_2 &= 0.5 & \overline{X}_2 &= 1.0289 \\
\beta_3 &= 0.51 & \overline{X}_3 &= 0.9711 \\
\beta_4 &= 0.5 & \overline{X}_4 &= 1.119 \\
\beta_5 &= 0.5 & \overline{X}_5 &= 0.881
\end{aligned}
$$

Inspection of Figures 7.1 and 7.2 reveals that the design is completely satisfactory. For example, in the case of the first pad, the approximated gap is reduced to $\overline{X}_1 = 0.709$. The stiffness will actually be increased at this gap and so the deflection is overestimated. The gap values for the other pads are all sufficiently close to the design values to avoid any concern.

■

Appendix: Tabular Design Procedures

Procedure 7.A1 Single Hydrostatic Pad

Step	Symbol	Description of Operation	Notes
1	W	Specify: normal load	From design requirements
2	W'	Specify: extreme load	From design requirements
3	P_s	Specify: supply pressure	Suggestion: 2 MN/m^2 (291 lbf/in^2)
4	β	Specify: pressure ratio, P_{ro}/P_s	Suggestion: 0.5 or see Chapter 7
5		Specify: land-width ratios	Choose optimum values from Chapter 4
6		Specify: land-width ratios	
7	$\overline{A}$	Read: area factor	See Chapter 4 for chosen pad type
8	$\overline{B}$	Read: flow factor	See Chapter 4 for chosen pad type
9	A	Calculate: pad area $A = W/(P_s\beta\overline{A})$	If projected area A is too large, increase P_s
10		Calculate: first pad dimension	From A calculate dimension
11	a	Calculate: land width	From land-width ratio
12		Calculate: second pad dimension	From A calculate dimension
13	h_o	Specify: bearing gap	More than 5–10 times flatness tolerance
14	U	Specify: sliding speed	If $U = 0$ or is very small, go to step 17 and specify η
15	A_r	Calculate: total recess area	From recess dimensions
16	A_f	Calculate: friction area	From $A_f = A - {}^3\!/_4 A_r$
17	η	Specify or calculate: viscosity	From $\eta = \dfrac{P_s h_o^2}{U}\sqrt{\dfrac{\beta\overline{B}}{A_f}}$
18	q_o	Calculate: flow	$q_o = P_s h_o^3 \beta\overline{B}/\eta$
19	$\overline{\lambda}$	Read: stiffness factor	From Figures 7.1–7.9
20	λ	Calculate: stiffness	From $P_s A \overline{A}\,\overline{\lambda}/h_o$
21	$\overline{W}'$	Calculate: extreme load factor	From $\overline{W}' = W'/(P_s A \overline{A})$
22	$\overline{X}$	Read: extreme gap factor	From Figures 7.1–7.8 using value of $\overline{W}'$
23	h'	Calculate: extreme gap	From $h' = h_o \overline{X}$
24	H_p	Calculate: pumping power	From $H_p = P_s q_o$
25	H_f	Calculate: friction power	From $H_f = \eta A_f U^2/h_o$
26	K	Calculate: power ratio	From $K = H_f/H_p$
27	ΔT	Calculate: max temp rise	SI units: $\Delta T = 0.6 \times 10^{-6} P_s (1 + K)$ °C

Example 7.A1 *Single Circular Hydrostatic Pad, Capillary Control, Speed $U = 0.5$ m/s, Normal Load 5000 N, Maximum Load 7000 N; Refer to Chapter 4 for Circular Pad Factors and This Chapter for Stiffness Factor*

Step	Symbol	Example of Working	Result
1	W	Normal load	5000 N (1124 lbf)
2	W'	Extreme load	7000 N (1574 lbf)
3	P_s	Supply pressure	2 MN/m^2 (291 lbf/in^2)
4	β	Pressure ratio	0.5
5	a/D	Land-width ratio	0.25
6			
7	$\overline{A}$	Area factor	0.541
8	$\overline{B}$	Flow factor	0.755
9	A	$\dfrac{5000}{2 \times 10^6 \times 0.5 \times 0.54}$	0.009259 m^2 (14.35 in^2)
10	$D = 2R_2$	$\sqrt{4 \times 0.009259/\pi}$	0.1086 m (4.274 in)
11	a	0.1086×0.25	0.0271 m (1.069 in)
12	R_1	$0.1086/2 - 0.0271$	0.0271 m (1.069 in)
13	h_o	Bearing gap	50 μm (0.002 in)
14	U	Sliding speed	0.5 m/s (19.68 in/s)
15	A_r	$\pi \times 0.0271^2$	0.002307 m^2 (3.576 in^2)
16	A_f	$0.009259 - 0.75 \times 0.002307$	0.007529 m^2 (11.67 in^2)
17	η	$\dfrac{2 \times 10^6 \times 50^2 \times 10^{-12}}{0.5} \sqrt{\dfrac{0.5 \times 0.755}{0.007529}}$	0.0708 N s/m^2 (70.8 cP)
18	q_o	$2 \times 10^6 \times 50^3 \times 10^{-18} \times 0.5 \times 0.755/0.0708$	1.33×10^{-6} m^3/s (0.08 l/min)
19	$\overline{\lambda}$	For value $\overline{X} = 1$	0.75
20	λ	$2 \times 10^6 \times 0.009259 \times 0.541 \times 0.75/50 \times 10^{-6}$	150 MN/m (0.86×10^6 lbf/in)
21	$\overline{W}'$	$7000/(2 \times 10^6 \times 0.009259 \times 0.541)$	0.6987
22	$\overline{X}'$	For $\overline{W}' = 0.6987$	0.75
23	h'	$50 \times 10^{-6} \times 0.75$	37.5 μm (0.0015 in)
24	H_p	$2 \times 10^6 \times 1.33 \times 10^{-6}$	2.66 W
25	H_f	$0.0708 \times 0.007529 \times 0.5^2/50 \times 10^{-6}$	2.66 W
26	K	$2.66/2.66$	1.0
27	ΔT	$0.6 \times 10^{-6} \times (1 + 1) \times 2 \times 10^6$	2.4 °C

Procedure 7.A2 Single Aerostatic Pad with Virtual Recess

Step	Symbol	Description of Operation	Notes
1	W	Specify: normal load $\times$ 1.15	From design requirements
2	W'	Specify: extreme load $\times$ 1.15	From design requirements
3	P_s	Specify: gauge supply pressure	Suggestion: 0.45 MN/m^2 (65 lbf/in^2)
4	K_{go}	Specify pressure ratio P_{ro}/P_s	Suggestion: 0.5 or see Chapter 7
5	p_{ro}	Calculate: absolute entry pressure	$p_{ro} = K_{go}P_s + p_a$
6		Specify: land-width ratio	Choose optimum values from Chapter 4
7	$\overline{A}$	Read or calculate	See Chapter 4 for chosen pad type
8	$\overline{B}$	Read or calculate	See Chapter 4 for chosen pad type
9	A	Calculate: $A = W/(P_s\beta\overline{A})$	If projected area A is too large, increase P_s
10		Calculate: first pad dimension	From A calculate pad dimension
11		Calculate: virtual recess sizes	For example: $D_r = (1 - 2a/D)D$ for circular pad
12	C_{vr}	Calculate: virtual recess perimeter	From step 11
13	n	Sources per virtual recess	
14	d_r	Specify: entry recess or slot width	Suggestion: $d_r \geq 0.7C_{vr}/n$
15	h_o	Specify: bearing gap	h_o should be 5–10 times flatness tolerance
16	η	Specify: viscosity for gas	For air $\eta = 0.0000183$ N s/m^2 at 18 °C
17	q_o	Calculate: free air flow with 70% virtual recess surround	$q_o \approx 0.85 \dfrac{P_s h_o^3 K_{go}\overline{B}}{\eta}\left(\dfrac{p_{ro} + p_a}{2p_a}\right)$
18	$\overline{\lambda}$	Read:	From Figures 7.1–7.9
19	λ	Calculate:	From $P_s A\overline{A}\,\overline{\lambda}/h_o$
20	$\overline{W}'$	Calculate: extreme load factor	$\overline{W}' = W'/(P_s A\overline{A})$
21	$\overline{X}$	Read: displacement factor	Refer Figures 7.1–7.8 for $\overline{W}'$
22	h'	Calculate: extreme gap	From $h' = h_o/\overline{X}$
23	H_p	Calculate: pumping power	From $H_p = p_a q_a \ln(p_s/p_a)$
24	H_f	Calculate: friction power	From $H_f = \eta AU^2/h_o$

Example 7.A2 Single Circular Aerostatic Pad, Orifice Control, Speed $U = 0.5$ m/s, Normal Load 5000 N, Maximum Load 7000 N

Step	Symbol	Example of Working	Result
1	W	Normal load × 1.15	5750 N (1293 lbf)
2	W'	Extreme load × 1.15	8050 N (1810 lbf)
3	P_s	Supply pressure	0.45 MN/m^2 (65 lbf/in^2)
4	K_{go}	P_{ro}/P_s	0.5
5	p_{ro}	$(0.5 \times 0.45 + 0.101) \times 10^6$	0.326 MN/m^2
6	a/D	Land-width ratio	0.25
7	$\overline{A}$	Area factor	0.541
8	$\overline{B}$	Flow factor	0.755
9	A	$\dfrac{5750}{0.45 \times 10^6 \times 0.5 \times 0.541}$	0.04724 m^2 (73.22 in^2)
10	D	$\sqrt{4 \times 0.04724/\pi}$	0.2453 m (9.66 in)
11	D_r	$(1 - 2 \times 0.25) \times 0.2453$	0.1227 m (4.83 in)
12	C_{vr}	$\pi \times 0.1227$	0.3855 m (15.18 in)
13	n		8
14	d_r	$d_r \geq 0.7 \times 0.3855/8$	0.03373 m (1.33 in) slot width
15	h_o	Bearing gap	25 μm (0.001 in)
16	η	Viscosity of air	18.3×10^{-6} N s/m^2
17	q_0	$\dfrac{0.85 \times 0.45 \times 10^6 \times 25^3 \times 10^{-18} \times 0.5 \times 0.755}{18.3 \times 10^{-6}}$ $\times \dfrac{0.326 + 0.101}{2 \times 0.101}$	0.0002606 m^3/s (15.6 l/min)
18	$\overline{\lambda}$	For $\overline{X} = 1$	0.65
19	λ	$0.45 \times 10^6 \times 0.04724 \times 0.541 \times 0.65/25 \times 10^{-6}$	299 MN/m (1.71×10^6 lbf/in)
20	$\overline{W}'$	$8050/(0.45 \times 10^6 \times 0.0411 \times 0.541)$	0.8045
21	$\overline{X}$	For $\overline{W}' = 0.8045$	0.5
22	h'	$25 \times 10^{-6} \times 0.5$	12.5 μm (0.0005 in)
23	H_p	$0.101 \times 10^6 \times 0.0002606 \times \ln(0.551/0.101)$	44.7 W
24	H_f	$0.0000183 \times 0.04724 \times 0.5^2/25 \times 10^{-6}$	0.0088 W

Procedure 7.A3 Equal Opposed Hydrostatic Pads

Step	Symbol	Description of Operation	Notes
1	W'	Specify: extreme load	From design requirements
2	P_s	Specify: supply pressure	Suggestion: 2 MN/m^2 (291 lbf/in^2)
3	β	Specify: pressure ratio, P_{ro}/P_s	Suggestion: 0.5 or see Chapter 7
4		Specify: land-width ratios	Choose optimum values from Chapter 4
5		Specify: land-width ratios	
6	$\overline{A}$	Read: area factor	See Chapter 4 for chosen pad type
7	$\overline{B}$	Read: flow factor	See Chapter 4 for chosen pad type
8	$\overline{W}$	Read: load factor	See Chapter 7 for acceptable value
9	A	Calculate: pad area $A = W/P_s\overline{A}\,\overline{W}$	If projected area A is too large, increase P_s
10		Calculate: first pad dimension	From A calculate dimension
11	a	Calculate: land width	From land-width ratio
12		Calculate: second pad dimension	From A calculate dimension
13	h_o	Specify: bearing gap	h_o should be at least 5–10 times flatness tolerance
14	U	Specify: sliding speed	If $U = 0$ or is very small, go to step 17 and specify η
15	A_r	Calculate: total recess area	From recess dimensions
16	A_f	Calculate: total friction area	From $A_f = A - {}^3\!/_4 A_r$
17	η	Specify or calculate: viscosity	From $\eta = \dfrac{P_s h_o^2}{U}\sqrt{\dfrac{2\beta\overline{B}}{A_f}}$
18	q_o	Calculate: flow	$q_o = 2P_s h_o^3 \beta\overline{B}/\eta$
19	$\overline{\lambda}$	Read: stiffness factor	From Figures 7.1–7.9
20	λ	Calculate: stiffness	From $P_s A\overline{A}\,\overline{\lambda}/h_o$
21	H_p	Calculate: pumping power	From $H_p = P_s q_o$ for two pads
22	H_f	Calculate: friction power	From $H_f = \eta A_f U^2/h_o$
23	K	Calculate: power ratio	From $K = H_f/H_p$
24	ΔT	Calculate: max temp rise	SI units: $\Delta T = 0.6 \times 10^{-6} P_s (1 + K)$ °C

Example 7.A3 Square, Opposed Plane Hydrostatic Pads, Capillary Control, Sliding Speed 0.5 m/s, Maximum Applied Load 7000 N

Step	Symbol	Example of Working	Result
1	W'	Extreme load	7000 N (1574 lbf)
2	P_s	Supply pressure	2 MN/m^2 (291 lbf/in^2)
3	β	Pressure ratio	0.5
4	a/L	Land-width ratio	0.25
5			
6	$\overline{A}$	Area factor	0.525
7	$\overline{B}$	Flow factor	0.85
8	$\overline{W}$	Load factor from Figure 7.10	0.4
9	A	$\dfrac{7000}{2 \times 10^6 \times 0.5 \times 0.525 \times 0.4}$	0.03333 m^2 (51.66 in^2)
10	L	$\sqrt{0.03333/2}$	0.1291 m (5.083 in)
11	a	0.1291×0.25	0.03225 m (1.269 in)
12			
13	h_o	Bearing gap	25 μm (0.001 in)
14	U	Sliding speed	0.5 m/s (19.68 in/s)
15	A_r	$2 \times (0.1291 - 2 \times 0.3225)^2$	0.008346 m^2 (12.94 in^2)
16	A_f	$0.03333 - 0.75 \times 0.008346$	0.02707 m^2 (41.96 in^2)
17	η	$\dfrac{2 \times 10^6 \times 25^2 \times 10^{-12}}{0.5} \sqrt{\dfrac{2 \times 0.5 \times 0.85}{0.02707}}$	0.01401 N s/m^2 (14.01 cP)
18	q_o	$\dfrac{2 \times 2 \times 10^6 \times 25^3 \times 10^{-18} \times 0.5 \times 0.85}{0.01401}$	1.896×10^{-6} m^3/s (0.114 l/min)
19	$\overline{\lambda}$	For value $\overline{X} = 1$	0.75
20	λ	$\dfrac{2 \times 10^6 \times 0.01666 \times 0.525 \times 0.75}{25 \times 10^{-6}}$	525 MN/m (3×10^6 lbf/in)
21	H_p	$2 \times 10^6 \times 2.681 \times 10^{-6}$	3.792 W
22	H_f	$0.01401 \times 0.02707 \times 0.5^2/25 \times 10^{-6}$	3.792 W
23	K	$3.792/3.792$	1.0
24	ΔT	$0.6 \times 10^{-6} \times (1+1) \times 2 \times 10^6$	2.4 °C

Procedure 7.A4 Equal Opposed Aerostatic Pads

Step	Symbol	Description of Operation	Notes
1	W'	Specify: extreme load × 1.15	From design requirements
2	P_s	Specify: supply pressure	Suggestion: $P_s = 0.45$ MN/m^2 (65 lbf/in^2)
3	K_{go}	Specify: pressure ratio, P_{ro}/P_s	Suggestion: 0.5 or see Chapter 7
4	p_{ro}	Calculate: absolute entry pressure	From $K_{go}P_s + p_a$
5		Specify: land-width ratios	Choose optimum values from Chapter 4
6		Specify: land-width ratios	
7	$\overline{A}$	Read: area factor	See Chapter 4 for chosen pad type
8	$\overline{B}$	Read: flow factor	See Chapter 4 for chosen pad type
9	$\overline{W}$	Read: load factor	See charts for acceptable value
10	A	Calculate: pad area $A = W/P_s\overline{A}\,\overline{W}$	If projected area A is too large, increase P_s
11		Calculate: first pad dimension	From A calculate dimension
12	a	Calculate: land width	From land-width ratio
13	C_{vr}	Calculate: virtual recess perimeter	From A calculate dimension
14	n	Specify: number of sources or slots	Suggestion: equi-spaced
15	d_r	Specify: entry recess or slot width	Suggestion: $d_r \geq 0.7 C_{vr}/n$
16	h_o	Specify: bearing gap	h_o should be at least 5–10 times flatness tolerance
17	η	Specify: viscosity of gas	For air $\eta = 0.0000183$ N s/m^2 at 18 °C
18	q_o	Calculate: flow with virtual recess	$q_o \approx 0.85 \times \dfrac{2P_s h_o^3 K_{go}\overline{B}}{\eta} \times \dfrac{p_{ro}+p_a}{2p_a}$
19	$\overline{\lambda}$	Read: stiffness factor	From Figures 7.1–7.9
20	λ	Calculate: stiffness	From $P_s A\overline{A}\,\overline{\lambda}/h_o$
21	H_p	Calculate: pumping power	From $H_p = p_a q_a \ln(p_s/p_a)$ for two pads
22	H_f	Calculate: friction power	From $H_f = \eta A_f U^2/h_o$

Example 7.A4 Square, Opposed Aerostatic Pads, Orifice Control, Sliding Speed 0.5 m/s, Maximum Applied Load 7000 N

Step	Symbol	Example of Working	Result
1	W'	7000×1.15	8050 N (1810 lbf)
2	P_s, p_s	Gauge and absolute supply pressure	$P_s = 0.45 \text{ MN/m}^2$
			$p_s = 0.551 \text{ MN/m}^2$
3	K_{go}	Pressure ratio	0.5
4	p_{ro}	$(0.5 \times 0.45 + 0.101) \times 10^6$	0.326 MN/m^2
5	a/L	Land-width ratio	0.25
6			
7	$\overline{A}$	Area factor	0.525
8	$\overline{B}$	Flow factor	0.85
9	$\overline{W}$	Load factor from Figure 7.12	0.4
10	A	$\dfrac{8050}{2 \times 10^6 \times 0.5 \times 0.525 \times 0.4}$	0.03833 m^2 (59.42 in^2)
11	L	$\sqrt{0.03833/2}$	0.1384 m (5.45 in)
12	a	0.1384×0.25	0.0346 m (1.362 in)
13	C_{vr}	$(0.1384 - 2 \times 0.0346) \times 4$	0.2768 m (10.9 in)
14	n	Multiple of 4 for a square pad	8
15	d_r	$d_r \geq 0.7 \times 0.2768/8$	0.02422 m (0.9535 in) slot width
16	h_o	Bearing gap	25 µm (0.001 in)
17	η	Viscosity of air	$18.3 \times 10^{-6} \text{ N s/m}^2$
18	q_o	$\dfrac{1.7 \times 0.45 \times 10^6 \times 25^3 \times 10^{-18} \times 0.5 \times 0.85}{0.0000183}$ $\dfrac{0.326 + 0.101}{2 \times 0.101}$	$587 \times 10^{-6} \text{ m}^3/\text{s}$ (35.2 l/min)
19	$\overline{\lambda}$	For value $\overline{X} = 1$	1.0
20	λ	$\dfrac{0.45 \times 10^6 \times 0.03833 \times 0.525 \times 1.0}{25 \times 10^{-6}}$	362 MN/m $(2.07 \times 10^6 \text{ lbf/in})$
21	H_p	$0.101 \times 10^6 \times 587 \times 10^{-6} \times \ln(0.551/0101)$	101 W
22	H_f	$18.3 \times 10^{-6} \times 0.03833 \times 0.5^2/25 \times 10^{-6}$	0.007 W

Procedure 7.A5 Unequal Opposed Hydrostatic Pads

Step	Symbol	Description of Operation	Notes
1	W_{pos}	Specify: maximum positive load	From design requirements
2	W_{neg}	Specify: maximum negative load	From design requirements
3	E	Read from Figure 7.13	Area ratio $E = A_{e1}/A_{e2}$
4	P_s	Specify: supply pressure	Suggestion: 2 MN/m^2 (291 lbf/in^2)
5		Specify: land-width ratios	Choose optimum values from Chapter 4
6		Specify: land-width ratios	
7	$\overline{A}$	Read: area factor	See Chapter 4 for chosen pad type
8	$\overline{B}$	Read: flow factor	See Chapter 4 for chosen pad type
9	$\overline{W}_{pos}$	Read: load factor	See Figure 7.14
10	$\overline{W}_{neg}$	Read: load factor	See Figure 7.14
11	A_{min}	Calculate: pad area $A = W_{pos}/P_s\overline{A}\,\overline{W}$	$A_{min} = A_1 + A_2$
12	A_{min}	Calculate: pad area $A = W_{neg}/P_s\overline{A}\,\overline{W}$	
13	A_{actual}	Specify: convenient value	
14	A_1	Calculate	$A_1 = A \cdot E/(1+E)$
15	A_2	Calculate	$A_2 = A/(1+E)$
16		Calculate: pad dimensions	
17		Calculate: pad dimensions	
18		Calculate: pad dimensions	
19		Calculate: pad dimensions	
20		Calculate: pad dimensions	
21		Calculate: pad dimensions	
22	h_o	Specify: bearing gap (single)	h_o should be at least 5–10 times flatness tolerance
23	U	Specify: sliding speed	If $U = 0$ or is very small, go to step 25 and specify η
24	A_r	Calculate: total recess area	$A_r = A_{r1} + A_{r2}$
25	A_f	Calculate: total friction area	From $A_f = A - {}^3\!/_4 A_r$
26	η	Specify or calculate: viscosity	From $\eta = \dfrac{P_s h_o^2}{U}\sqrt{\dfrac{2\beta\overline{B}}{A_f}}$
27	q_o	Calculate: flow	$q_o = 2P_s h_o^3 \beta\overline{B}/\eta$
28	$\overline{\lambda}$	Read: stiffness factor	From Figure 7.15
29	λ	Calculate: stiffness	From $P_s A \overline{A}\overline{\lambda}/h_o$
30	H_p	Calculate: pumping power	From $H_p = P_s q_o$ for two pads
31	H_f	Calculate: friction power	From $H_f = \eta A_f U^2/h_o$
32	K	Calculate: power ratio	From $K = H_f/H_p$
33	ΔT	Calculate: max temp rise per pass	SI units: $\Delta T = 0.6 \times 10^{-6} P_s (1+K)$ °C

Example 7.A5 Unequal Opposed Hydrostatic Pads: Capillary Control, Sliding Speed 1.27 m/s, Maximum Load 8900 N, Design Pressure Ratio 0.5, Rectangular Pads 2:1 Length/Width

Step	Symbol	Example of Working	Result
1	W_{pos}	Maximum positive load	8900 N (2000 lbf)
2	W_{neg}	Maximum negative load	2450 N (550 lbf)
3	E	Approximate value of 2	2
4	P_s	Supply pressure	2 MN/m^2 (291 lbf/in^2)
5	a/L	Land-width ratio	0.25
6			
7	$\overline{A}$	Area factor	0.63
8	$\overline{B}$	Flow factor	1.45
9	$\overline{W}_{pos}$	Load factor from Figure 7.14	0.5
10	$\overline{W}_{neg}$	Load factor from Figure 7.14	0.14
11	A_{min}	$8900/(2 \times 10^6 \times 0.63 \times 0.5)$	0.0141 m^2 (21.9 in^2)
12	A_{min}	$2450/(2 \times 10^6 \times 0.63 \times 0.14)$	0.014 m^2 (21.7 in^2)
13	A_{actual}	Convenient value	0.0144 m^2 (22.32 in^2)
14	A_1	$0.0144 \times 2/3$	0.0096 m^2 (14.88 in^2)
15	A_2	$0.0144/3$	0.0048 m^2 (7.44 in^2)
16	L_1	$\sqrt{0.0096/2}$	0.0693 m (2.73 in)
17	B_1	2×0.0693	0.1386 m (5.46 in)
18	a_1	0.0693×0.25	0.0173 m (0.6732 in)
19	L_2	$\sqrt{0.0048/2}$	0.049 m (1.93 in)
20	B_2	2×0.049	0.098 m (3.86 in)
21	a_2	0.049×0.25	0.01225 m (0.482 in)
22	h_o	Bearing gap	50 μm (0.002 in)
23	U	Sliding speed	1.27 m/s (50 in/s)
24	A_r	$0.00361 + 0.018$	0.0054 m^2 (8.4 in^2)
25	A_f	$0.0096 + 0.0048 - 0.75 \times 0.0054$	0.0103 m^2 (16.04 in^2)
26	η	$\dfrac{2 \times 10^6 \times 50^2 \times 10^{-12}}{1.27} \sqrt{\dfrac{2 \times 0.5 \times 1.45}{0.0103}}$	0.047 N s/m^2 (47 cP)
27	q_o	$\dfrac{2 \times 2 \times 10^6 \times 50^3 \times 10^{-18} \times 0.5 \times 1.45}{0.047}$	7.71×10^{-6} m^3/s (0.463 l/min)
28	$\overline{\lambda}$	For value $\overline{X} = 1$	0.75
29	λ	$2 \times 10^6 \times 0.0144 \times 0.63 \times 0.75/50 \times 10^{-6}$	272 MN/m (1.55×10^6 lbf/in)
30	H_p	$2 \times 10^6 \times 7.71 \times 10^{-6}$	15.4 W
31	H_f	$0.047 \times 0.0103 \times 1.27^2/50 \times 10^{-6}$	15.6 W
32	K	$15.4/15.6$	1.0
33	ΔT	$0.6 \times 10^{-6} \times 2 \times 10^6 \times (1+1)$	2.4 °C

Procedure 7.A6 Unequal Opposed Aerostatic Pads

Step	Symbol	Description of Operation	Notes
1	W_{pos}	Specify: maximum positive load	From design requirements
2	W_{neg}	Specify: maximum negative load	From design requirements
3	E	Read from Figure 7.13	Area ratio $E = A_{e1}/A_{e2}$
4	P_s	Specify: gauge supply pressure	Suggestion: $P_s = 0.45$ MN/m^2
5	K_{go}	Specify: pressure ratio, P_{ro}/P_s	Suggestion: 0.5
6	p_{ro}	Calculate: absolute entry pressure	From $K_{go}P_s + p_a$
7		Specify: land-width ratios	Choose optimum values from Chapter 4
8	$\overline{A}$	Read: area factor	See Chapter 4 for chosen pad type
9	$\overline{B}$	Read: flow factor	See Chapter 4 for chosen pad type
10	$\overline{W}_{pos}$	Read: load factor	See Figure 7.14
11	$\overline{W}_{neg}$	Read: load factor	See Figure 7.14
12	A_{min}	Calculate: pad area $A = W_{pos}/P_s\overline{A}\,\overline{W}$	$A_{min} = A_1 + A_2$
13	A_{min}	Calculate: pad area $A = W_{neg}/P_s\overline{A}\,\overline{W}$	
14	A_{actual}	Specify: convenient value	
15	A_1	Calculate	$A_1 = A \cdot E/(1+E)$
16	A_2	Calculate	$A_2 = A/(1+E)$
17		Calculate: pad dimensions	
18		Calculate: pad dimensions	
19		Calculate: pad dimensions	
20		Calculate: pad dimensions	
21		Calculate: pad dimensions	
22	C_{vr}	Calculate: virtual recess perimeter	Also calculate for small virtual recess
23	n	Specify: number of entry sources	Suggestion: equi-spaced
24	d_r	Specify: source width	Suggestion: $d_r \geq 0.7C_{vr}/n$
25	h_o	Specify: bearing gap	h_o should be at least 5–10 times flatness tolerance
26	η	Specify: viscosity of gas	For air $\eta = 0.0000183$ N s/m^2 at 18 °C
27	q_o	Calculate: flow for virtual recess	$q_o \approx 0.85 \times \dfrac{2P_s h_o^3 K_{go}\overline{B}}{\eta} \times \dfrac{p_{ro} + p_a}{2p_a}$
28	$\overline{\lambda}$	Read: stiffness factor	From Figures 7.1–7.9
29	λ	Calculate: stiffness	From $P_s A\overline{A}\,\overline{\lambda}/h_o$
30	H_p	Calculate: pumping power	From $H_p = p_a q_a \ln(p_s/p_a)$ for two pads
31	H_f	Calculate: friction power	From $H_f = \eta A_f U^2/h_o$

Example 7.A6 Square, Unequal Opposed Aerostatic Pads, Orifice Controlled, Sliding Speed 0.5 m/s, Maximum Applied Load 4500 N, Ambient Pressure 0.101 MN/m^2

Step	Symbol	Example of Working	Result
1	W_{pos}	Maximum positive load $\times$ 1.15	5175 N (1163 lbf)
2	W_{neg}	Maximum negative load $\times$ 1.15	1450 N (346 lbf)
3	E	Approximate value of 2	2
4	P_s	Gauge supply pressure	$P_s = 0.45$ MN/m^2
5	K_{go}	Gauge pressure ratio	0.5
6	p_{ro}	$(0.5 \times 0.45 + 0.101) \times 10^6$	0.326 MN/m^2
7	a/L	Land-width ratio	0.25
8	$\overline{A}$	Area factor	0.525
9	$\overline{B}$	Flow factor	0.85
10	$\overline{W}_{pos}$	Load factor from Figure 7.14	0.5
11	$\overline{W}_{neg}$	Load factor from Figure 7.14	0.14
12	A_{min}	$5175/(0.45 \times 10^6 \times 0.525 \times 0.5)$	0.0438 m^2 (67.9 in^2)
13	A_{min}	$1450/(0.45 \times 10^6 \times 0.525 \times 0.14)$	0.0438 m^2 (67.9 in^2)
14	A_{actual}	Convenient value	0.044 m^2 (68.2 in^2)
15	A_1	$0.044 \times 2/3$	0.02933 m^2 (45.5 in^2)
16	A_2	$0.044/3$	0.01467 m^2 (22.73 in^2)
17	L_1	$\sqrt{0.02933}$	0.1713 m (6.743in)
18	a_1	0.1713×0.25	0.0428 m (1.686 in)
19	L_2	$\sqrt{0.01467}$	0.1211 m (4.768 in)
20	a_2	0.1211×0.25	0.03028 m (1.192 in)
21			
22	C_{vr}	$(0.1713 - 2 \times 0.0428) \times 4$	0.3426 m (13.66 in) and 0.2422 m
23	n	Multiple of 4 for square pad	8
24	d_r	$d_r \geq 0.7 \times 0.3428/8$	0.03 m (1.18 in) and 0.0212 m (0.835 in)
25	h_o	Bearing gap	25 μm (0.001 in)
26	η	Viscosity of air	18.3×10^{-6} N s/m^2
27	q_o	$\dfrac{1.7 \times 0.45 \times 10^6 \times 25^3 \times 10^{-18} \times 0.5 \times 0.85}{0.0000183}$ $\dfrac{0.326 + 0.101}{2 \times 0.101}$	587×10^{-6} m^3/s (35.22 l/min)
28		For value $\overline{X} = 1$	1.0
29	λ	$0.45 \times 10^6 \times 0.044 \times 0.525 \times 1.0/25 \times 10^6$	416 MN/m (2.38×10^6 lbf/in)
30	H_p	$0.101 \times 587 \times \ln(0.551/0.101)$	101 W
31	H_f	$18.3 \times 10^{-6} \times 0.044 \times 0.5^2/25 \times 10^6$	0.0081 W

CHAPTER 8

Partial Journal Bearings

Summary of Key Design Formulae

$$q = \frac{P_r h_o^3}{\eta} \cdot \overline{B}$$ Hydrostatic flow

$$q = \frac{P_r h_o^3}{\eta} \cdot \overline{B} \cdot \left(\frac{p_r + p_a}{2 p_a} \right)$$ Aerostatic flow

$$W = P_r A \cdot \overline{A}$$ Load support

$$e = h_o \cdot \varepsilon$$ Journal eccentricity

8.1 Recessed Partial Journal Bearings

Hydrostatic bearings can be used to support a loaded shaft for both rotational and reciprocating motion. A full 360° journal bearing is unnecessary when the applied load is unidirectional and a partial pad may sometimes be used, as shown in Figure 8.1. These bearings are designed in a similar manner to plane bearings, as described in Chapter 7. Load and flow data are given in Figures 8.3–8.6. Combinations of partial bearings can also be designed to take side thrust as shown in Figure 8.2. Methods of combination are also as given for plane bearings in Chapter 7.

Aerostatic bearings employ very small grooves or virtual recesses to reduce recess volume instead of large recesses as used in hydrostatic bearings. This difference in recesses is important for dynamic stability, as described in previous chapters (Section 4.2).

Bearing projected area is $A = LD\sin(\phi/2)$, where ϕ typically varies between 80° and 180°. Figures 8.3–8.6 give values of thrust coefficient $\overline{A}$ and flow coefficient $\overline{B}$.

Land width, a, for short L/D pads is given in the form a/L. The land width must be the same in both the axial and the circumferential directions. However, as L/D increases, the recommended land-width ratio is applied for the circumferential land. Thus, with $L/D = 4$ and $\phi = 90°$ it is impracticable to write $a/L = 0.25$ since this would make the circumferential land width 1.27 times greater than circumferential pad length, that is $a/(\phi D/2) = 1.27$. The land-width ratio is therefore taken as the smaller of a/L or $2a/\phi D$. The following example illustrates the application of the charts.

Hydrostatic, Aerostatic and Hybrid Bearing Design.
DOI: 10.1016/B978-0-12-396994-1.00008-5

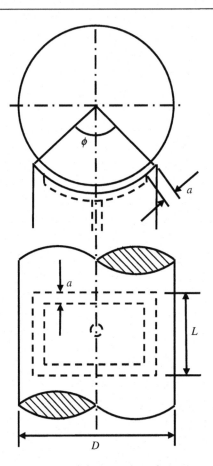

Figure 8.1: Partial Journal Pad Geometry.

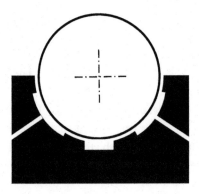

Figure 8.2: Partial Journal Bearing with Side Thrusts.

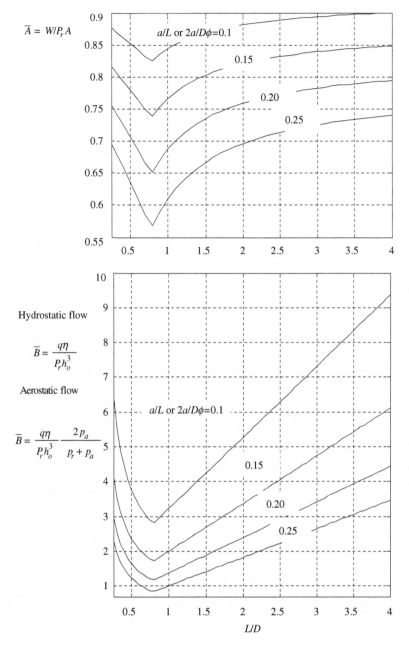

Figure 8.3: Area and Flow Coefficients for Partial Journal Pads, $\phi = 90°$.

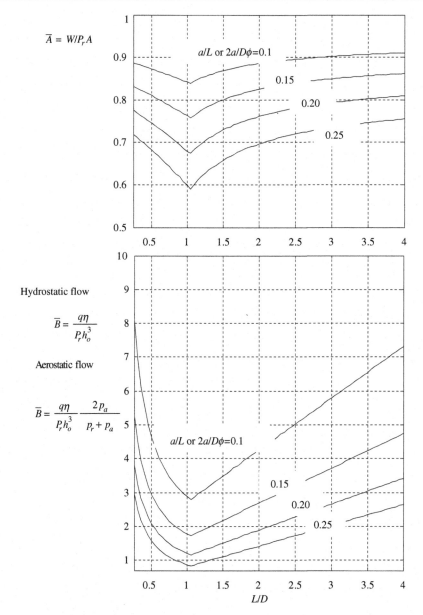

Figure 8.4: Area and Flow Coefficients for Partial Journal Pads, $\phi = 120°$.

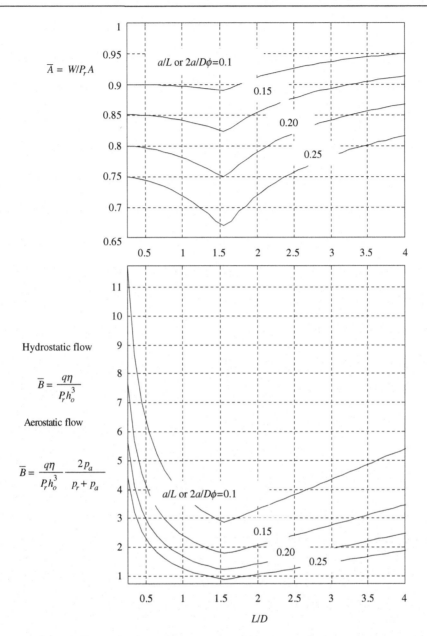

Figure 8.5: Area and Flow Coefficients for Partial Journal Pads, $\phi = 180°$.

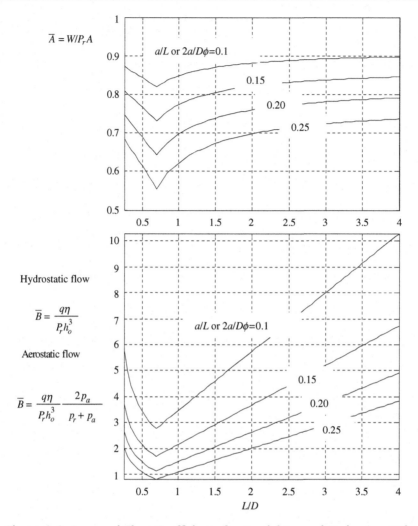

$$\overline{A} = W/P_r A$$

Hydrostatic flow

$$\overline{B} = \frac{q\eta}{P_r h_o^3}$$

Aerostatic flow

$$\overline{B} = \frac{q\eta}{P_r h_o^3} \frac{2p_a}{p_r + p_a}$$

Figure 8.6: Area and Flow Coefficients for Partial Journal Pads, $\phi = 80°$.

■ Example 8.1 Recessed Partial Hydrostatic Journal Bearing

Given data:

$\phi = 90°$
$D = 75$ mm (2.953 in)
$L = 60$ mm (2.362 in)
$A = 0.060 \times 0.075 \times \sin(90/2) = 0.003182$ m² (4.932 in)
$a = 15$ mm (0.5905 in)

$P_r = 3.5$ MN/m^2 (508 lbf/in^2)

$h_o = 25$ μm (0.001 in)

$\eta = 34.5$ cP (5×10^{-6} reyn).

From Figure 8.3, with $L/D = 60/75 = 0.8$ and $a/L = 15/60 = 0.25$,

$\overline{A} = 0.57$

$\overline{B} = 0.83$

$W = P_r A \overline{A} = 3.5 \times 10^6 \times 0.003182 \times 0.57 = 6.35$ kN (1427 lbf)

$$q = \frac{P_r h^3}{\eta} \overline{B} = \frac{3.5 \times 10^6 \times 25^3 \times 10^{-18}}{0.0345} \times 0.83 = 1.3 \times 10^{-6} \text{ m}^3/\text{s} \ (0.0793 \text{ in}^3/\text{s})$$

■

■ *Example 8.2 Partial Aerostatic Journal Bearing with a Virtual Recess*

Given data:

$\phi = 90°$

$D = 75$ mm (2.953 in)

$L = 60$ mm (2.362 in)

$A = 0.060 \times 0.075 \times \sin(90/2) = 0.003182$ m^2 (4.932 in)

$a = 15$ mm (0.5905 in)

$P_r = 0.25$ MN/m^2 (36.26 lbf/in^2)

$h_o = 25$ μm (0.001 in)

$\eta = 18.3 \times 10^{-6}$ N s/m^2 (0.002654 μ reyn).

From Figure 8.3, with $L/D = 60/75 = 0.8$ and $a/L = 15/60 = 0.25$,

$\overline{A} = 0.57$

$\overline{B} = 0.83$

$W = P_r A \overline{A} = 0.25 \times 10^6 \times 0.003182 \times 0.57 = 453$ N (102 lbf)

A virtual recess of 70% surround reduces thrust about 15% to 385 N.

$$q = \frac{P_r h^3}{\eta} \overline{B} \frac{p_r + p_a}{2p_a} = \frac{0.25 \times 10^6 \times 25^3 \times 10^{-18}}{18.3 \times 10^{-6}} \times 0.83 \times \frac{0.452}{0.202}$$

$$= 396 \times 10^{-6} \text{ m}^3/\text{s} \ (24 \text{ l/min})$$

A virtual recess with 70% surround reduces flow by about 15% to 337×10^{-6} m^3/s of free air.

An aerostatic partial journal bearing employing a virtual recess has an advantage compared with a similar bearing with an actual recess in that some side loading can be withstood.

■

Figure 8.2 shows a hydrostatic lift bearing designed to accommodate side thrust loading as well as gravity loading. Such a bearing can be designed using the data for two 80° arc single partial bearings and summing the vertical bearing load components. Suitable data for 80° pads are given in Figure 8.6. Thus, for two 80° pads (Figure 8.2), vertical lift can be found by multiplying the thrust for one pad obtained from Figure 8.6 by the factor $2 \times \sin 50° = 1.53$.

8.2 Partial Journal Bearings with a High-Pressure Supply Groove

A partial hydrostatic/hydrodynamic journal bearing has been described by Fuller (1956). Although hydrodynamic bearings behave satisfactorily when operating at the design speed, full-film lubrication does not exist under starting, reversing, and stopping conditions. Consequently wear, stick-slip vibration, and high friction may result. The problem can be overcome by use of a hydrostatic lift, essentially a high-pressure supply groove between the journal and the bearing where contact would occur (Figure 8.7).

Since the bearing must operate as a hydrodynamic bearing under normal load conditions, the hydrostatic groove or recess must be minimal to avoid interrupting the hydrodynamic "wedge" surfaces of the bearing.

This bearing is not the same as hybrid bearings, where the hydrostatic and hydrodynamic actions are provided continuously. In the grooved bearing considered here, hydrostatic lift is only employed under starting and stopping conditions. The bearing is usually supplied by a constant-flow pump capable of producing the pressure necessary to lift the shaft in its most adverse position when the shaft is in contact with the bearing. The bearing groove area must be large enough to support the load under maximum pressure of the pump. Fuller (1956) gives the following hydrostatic design equations for 180° bearings:

$$\overline{A} = \frac{b}{L} \frac{2 + 3\varepsilon - \varepsilon^3}{\varepsilon(4 - \varepsilon^3) + \frac{4 + 2\varepsilon^2}{\sqrt{1 - \varepsilon^2}} \arctan\left(\frac{1 + \varepsilon}{1 - \varepsilon^2}\right)} = \frac{b}{L} f_1(\varepsilon) \tag{8.1}$$

$$\overline{B} = \frac{4b}{D} \frac{(1 - \varepsilon^2)^2}{6\left[\varepsilon(4 - \varepsilon^2) + \frac{4 + 2\varepsilon^2}{\sqrt{1 - \varepsilon^2}} \arctan\left(\frac{1 + \varepsilon}{\sqrt{1 - \varepsilon^2}}\right)\right]} = \frac{4b}{D} f_2(\varepsilon) \tag{8.2}$$

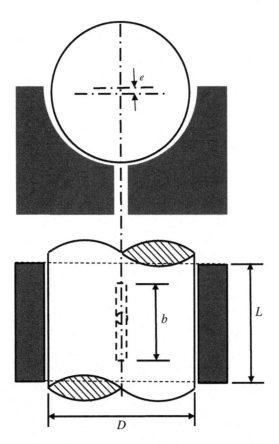

Figure 8.7: Partial Journal Bearing with High-Pressure Supply Groove.

where eccentricity ratio, ε, is given by

$$\varepsilon = e/h_o \tag{8.3}$$

Lift is given by

$$W = P_r L D \bar{A} \tag{8.4}$$

Flow is given by

$$q = \frac{P_r h_o^3}{\eta} \bar{B} \tag{8.5}$$

Figure 8.8 gives the variation of $f_1(\varepsilon)$ and $f_2(\varepsilon)$ with the eccentricity ratio ε. It is seen that support thrust reduces towards zero at touchdown when the eccentricity ratio is $\varepsilon = 1$. This means that sufficient pressure must be applied over the groove area alone to support the required initial lift.

The following example illustrates the method of design.

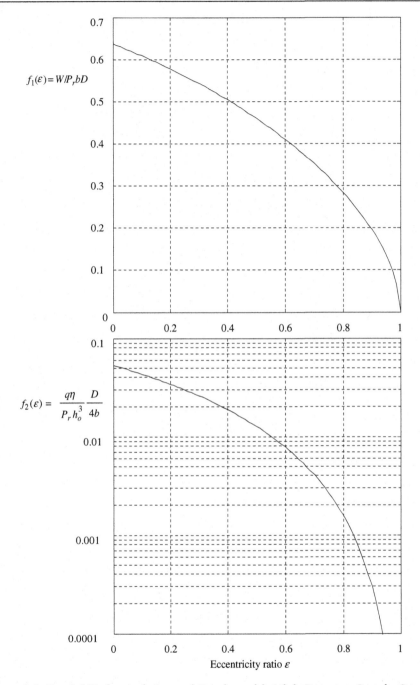

Figure 8.8: Partial Hydrostatic Journal Bearing with High-Pressure Supply Groove.

■ *Example 8.3 Grooved Partial Hydrostatic Journal Bearing*

With reference to Figure 8.7, given data are:

$D = 250$ mm (9.843 in)
$L = 300$ mm (11.81 in)
$h_o = 0.19$ mm (0.0075 in)
$W = 44$ kN (9890 lbf).

It is required to design a suitable hydrostatic lift for this bearing and determine the necessary pressure and flow at lifts of 25 and 60 μm using oil of dynamic viscosity 58 cP.

Solution

Try a groove $b/L = 0.5$, where $b = 150$ mm (5.91 in):

1. At $h = 25$ μm,

$$\varepsilon = \frac{e}{h_o} = \frac{h_o - h}{h_o} = \frac{0.19 - 0.025}{0.19} = 0.8684$$

From Figure 8.8,

$f_1(\varepsilon) = 0.22$, $b/L = 0.5$, and $\overline{A} = 0.5 \times 0.22 = 0.11$

$f_2(\varepsilon) = 0.00055$, $4b/D = 4 \times 150/250 = 2.4$, and $\overline{B} = 2.4 \times 0.00055 = 0.00132$

$$P_r = \frac{W}{LD\overline{A}} = \frac{44,000}{0.3 \times 0.25 \times 0.11} = 5.33 \text{ MN/m}^2 \ (773 \text{ lbf/in}^2)$$

$$q = \frac{P_r h_o^3}{\eta} \overline{B} = \frac{5.33 \times 10^6 \times 0.19^3 \times 10^{-9} \times 0.00132}{0.058}$$

$$= 0.83 \times 10^{-6} \text{ m}^3/\text{s} \ (0.051 \text{ in}^3/\text{s})$$

2. At $h = 60$ μm,

$$\varepsilon = \frac{h_o - h}{h_o} = \frac{0.19 - 0.06}{0.19} = 0.684$$

From Figure 8.8,

$f_1(\varepsilon) = 0.38$, $b/L = 0.5$, and $\overline{A} = 0.5 \times 0.38 = 0.19$

$f_2(\varepsilon) = 0.005$, $4b/D = 4 \times 150/250 = 2.4$, and $\overline{B} = 2.4 \times 0.005 = 0.012$

$P_r = 3.09 \text{ MN/m}^2 \ (448 \text{ lbf/in}^2)$

$q = 4.385 \times 10^{-6} \text{ m}^3/\text{s} \ (0.268 \text{ in}^3/\text{s})$

3. Assume the maximum pump pressure P_m for initial lift is 20 MN/m² (2900 lbf/in²),

$$\text{Groove width} = \frac{W}{P_m b} = \frac{44,000}{20 \times 10^6 \times 150 \times 10^{-3}} = 0.0147 \text{ m} \ (0.579 \text{ in})$$

The high initial value compared with the running value is typical. Fuller (1956) quotes a recess area 2.5–5% of the total projected area and a pump capacity five times the normal running pressure when the required lift is established.

■

Reference

Fuller, D. D. (1956). *Theory and Practice of Lubrication for Engineers*. New York: Wiley.

CHAPTER 9

Recessed Hydrostatic Journal Bearings

Summary of Key Design Formulae

$$q_o = \frac{P_s h_o^3}{\eta} \cdot \beta \cdot \overline{B}$$ Hydrostatic flow

$$W = P_s LD \cdot \overline{W}$$ Bearing film load support

$$\lambda = \frac{P_s LD}{h_o} \cdot \overline{\lambda}$$ Bearing film stiffness

$$K = \frac{H_f}{H_p}$$ Power ratio

9.1 Introduction

Application of Journal Bearings

Recessed journal bearings offer high rotational accuracy, high stiffness, and low temperature rise. Cylindrical bearings are the most common configuration. Alternatives include conical and spherical journal bearings, as described in Chapter 1. Recessed cylindrical bearings are a popular choice although plain bearings have advantages, as described in Chapter 10.

Two configurations are shown in Figure 9.1. The typical bearing is the simpler geometry without axial drainage grooves between the recesses. Axial drainage grooves reduce projected area and detract from hydrostatic and hydrodynamic load support. A possible advantage of axial grooves is greater through-flow for the same clearance.

Four, five, or six recesses are typical, recess depth typically being 20 times the gap at the bearing lands. Each recess is controlled by its own restrictor so that each recess with its surrounding land acts as a thrust pad; a whole bearing consists of *n* pads, of which *n*/2 pads support vertical loads and *n*/2 pads support horizontal loads.

Eccentricity Ratio

Journal bearing performance is complicated by non-uniform film thickness under load and the effect on recess pressures of inter-recess flows. Film thickness is shown greatly exaggerated in

Hydrostatic, Aerostatic and Hybrid Bearing Design.
DOI: 10.1016/B978-0-12-396994-1.00009-7
179

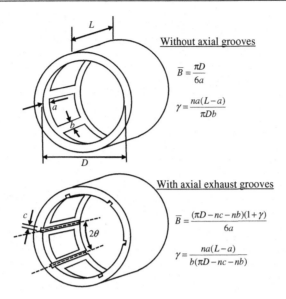

$$\underline{\text{Without axial grooves}}$$

$$\bar{B} = \frac{\pi D}{6a}$$

$$\gamma = \frac{na(L-a)}{\pi Db}$$

$$\underline{\text{With axial exhaust grooves}}$$

$$\bar{B} = \frac{(\pi D - nc - nb)(1+\gamma)}{6a}$$

$$\gamma = \frac{na(L-a)}{b(\pi D - nc - nb)}$$

Figure 9.1: Recessed Journal Bearings and Thin-Land Flow Factors.

Figure 9.2, where it is seen that film thickness h is related to eccentricity ratio ε and angular position ϕ from the line of eccentricity according to $h = h_o(1 - \varepsilon\cos\phi)$. Eccentricity ratio is defined as $\varepsilon = e/h_o$.

Recommended Design Values and Tolerances

A recommended design is a four-, five-, or six-recess bearing: $L/D = 1$ is recommended for length-to-diameter ratio; $a/L = 0.25$ is recommended for land-width ratio; and $\beta = 0.5$ is recommended for concentric pressure ratio. Capillary restrictors are recommended for flow control. In practice, the values adopted will vary to suit particular needs of a design. Figure 9.8, together with other charts, is designed to assist in the selection of a suitable combination of viscosity, gap, and supply pressure for a capillary-controlled bearing.

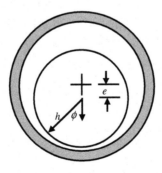

Figure 9.2: Film Thickness Under a Journal Bearing Land.

Sections 9.6 and 9.7, together with Example 9.4, illustrate the specification of a tolerance for bearing film gap. Bearing gap, in engineering practice, is usually termed "clearance". The method ensures a satisfactory operating range for both clearance and the associated concentric pressure ratio. Designers can include a tolerance in their designs using either the approximate design procedure or the exact method by adopting a simple practice. Maximum clearance is associated with minimum pressure ratio $\beta = 0.5$ in calculations. Minimum clearance is then specified to be approximately 0.7 times maximum clearance. This is usually stated conversely: maximum clearance should be larger than minimum clearance but not exceeding 1.5 times. These recommendations are explained in Chapter 6.

9.2 Flow

Flow In

Flow into each recess depends on the supply pressure and the required recess pressure. It also depends on the resistance of the restrictor for each recess, as explained in Section 2.6.

Flow Out

Thin-land expressions based on Figure 9.1 overestimate concentric flows if applied for thick-land bearings. The overestimate is less than 15% for $a/L = 0.25$ and $L/D < 2.0$. However, q increases with ε by as much as 50% at high values of pressure ratio. Provision should be made for maximum flow across the range of eccentricities and pressure ratios.

(i) Maximum Concentric Flow

Bearings do not need axial exhaust grooves and most bearings are designed without them. Bearings without axial grooves are more efficient. *Without axial grooves*, flow out of the bearing must all escape across the two end lands by traveling a distance a through the bearing gap h (see Figures 9.1 and 9.2). A generous estimate can be made of flow from equation (9.1) for a concentric shaft. Equation (9.1) gives maximum concentric flow q_o:

$$q_o = \frac{P_{ro}\overline{B}h_o^3}{\eta} \tag{9.1}$$

Each restrictor is designed to allow the required flow from a single recess. Maximum concentric flow from one recess is given by q_o divided by the number of recesses n:

$$q_{\text{restrictor}} = \frac{q_o}{n} \tag{9.2}$$

$$\overline{B} = \frac{\pi D}{6a} \tag{9.3}$$

where $P_{ro} = \beta P_s$.

(ii) Computed Data for Flow

Publications by O'Donoghue and Rowe (1968) and O'Donoghue et al. (1970a, b, 1971) describe finite difference techniques for computation of hydrostatic bearings. Data charts for $q\eta/P_s h_o^3$ are presented in Figure 9.5 based on finite difference computation taking into account shaft eccentricity. Flow is calculated from the pressure-flow field by a summation of the form:

$$q = 2x \int_0^{\pi D} \frac{h^3}{12\eta} \frac{dp}{dz} \, dx \tag{9.4}$$

Computed data are presented in Figure 9.5a–c in the form $q\eta/P_s h_o^3$.

Flow for the concentric case is given approximately by

$$\frac{q\eta}{P_s h_o^3} \approx \beta \overline{B} \tag{9.5}$$

Flow tends to increase with eccentricity ratio and more so at high values of pressure ratio.

■ Example 9.1 Flow

Given data in SI units for Bearing A and in British ips units for Bearing B:

Bearing A	Bearing B
$D = 50$ mm	$D = 2$ in
$L = 30$ mm	$L = 1$ in
$a = 6$ mm	$a = 0.25$ in
$h_o = 25 \ \mu$m	$h_o = 0.001$ in
$P_s = 2$ MN/m^2	$P_s = 300$ lbf/in^2
$\eta = 35$ cP	$\eta = 5 \times 10^{-6}$ reyn
$\beta = 0.5$	$\beta = 0.4$

Solution

Bearing A: L/D =0.6, a/L = 0.2, n = 4 recesses

For this case it is found that without interpolation, data are not available in Figure 9.5. Using the approximate expression, equation (9.5), for concentric flow gives

$$\beta \overline{B} = \beta \frac{\pi D}{6a} = 0.5 \times \frac{\pi.50}{6 \times 6} = 2.18$$

$$q_o = \frac{P_s h_o^3}{\eta} \beta \overline{B} = \frac{2 \times 10^6 \times 25^3 \times 10^{-18}}{0.035} \times 2.18 = 1.95 \times 10^{-6} \ \text{m}^3/\text{s} \ (0.117 \ \text{l/min})$$

Flow per restrictor $= q_o/n$, so that

$$q_{restrictor} = 1.95 \times 10^{-6}/4 = 0.488 \times 10^{-6} \ \text{m}^3/\text{s}$$

Bearing B: L/D = 0.5, a/L = 0.25, n = 4 recesses, and bridge angle θ = 10°
From Figure 9.5a, being careful to read the value for β = 0.4:

$$\frac{q\eta}{P_s h_o^3} = 1.66$$

$$q = \frac{300 \times 0.001^3}{5 \times 10^{-6}} \times 1.66 = 0.1 \ \text{in}^3/\text{s}$$

∎

Inter-Recess Flow and Bridge Angle

For an eccentric shaft, each recess pressure is different, as illustrated in Figure 1.3. Differences in recess pressures cause circumferential flow from one recess to another. Circumferential flow reduces bearing load support and also reduces film stiffness, particularly for $L/D > 1$.

The width, b, of the inter-recess land must therefore be large enough to prevent excessive circumferential flow. Applying equation (2.9), inter-recess flow from recess 1 to recess 2 may be estimated from

$$q_{1-2} = (P_1 - P_2)\frac{(L-a)h_{1-2}^3}{12\eta b} \tag{9.6}$$

Inter-recess land width b is often specified by bridge angle θ in the range $10° \leq \theta \leq 30°$ as in Figure 9.5a. Land widths are equal, $b = a$, when $L/D = 1$, $a/L = 0.25$, and $\theta = 28.6°$ or roughly 30°. The recommended value is 30° for a four-recess bearing.

Recirculatory Flow in a Recess

The lands surrounding the deep recesses of the bearing dam the flow giving rise to recirculation (Figure 9.3). At high speeds, recirculating flow increases friction power. This is particularly important with thin-land bearings, increasing the effective friction area of the recesses by

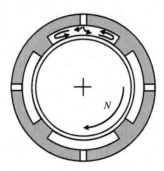

Figure 9.3: Recirculatory Flow in a Recess Increases Friction.

four or five times. The effect of recirculatory flow is taken into account when calculating friction power by modifying the expression for total friction area as follows:

$$A_f = A - \frac{3}{4}A_r \tag{9.7}$$

Turbulent and Vortex Flow

It is assumed for data charts that flow between bearing surfaces is simple laminar flow. At high speeds vortex flow commences and at even higher speeds flow becomes turbulent. The transition speed given by Taylor's criterion for onset of vortices is

$$\pi DN_t h \frac{\rho}{\eta} = 41.1\sqrt{\frac{D}{2h}} \tag{9.8}$$

where N_t is the transition speed.

■ Example 9.2 Transition Speed

A journal bearing, diameter 50 mm and radial clearance 75 μm, is to operate with oil of 34.5 cP viscosity at 38 °C. The oil density is 858 kg/m^3. What is the transition speed?

Solution
From equation (9.8):

$$\pi \times 50 \times 10^{-3} \times N_t \times 75 \times 10^{-6} \times \frac{858}{0.0345} = 41.1\sqrt{\frac{50 \times 10^{-3}}{2 \times 75 \times 10^{-6}}}$$

$$N_t = 2560 \text{ rev/s}$$

■

Hydrodynamic Flow

Journal speed causes additional velocity-induced circumferential flow from one recess to another, which helps to build up pressures under load. In thick-land bearings, a noticeable load is supported by additional hydrodynamic pressures at the lands. A hydrodynamic effect of journal speed is to cause the shaft to become eccentric at an angle ϕ in the direction of rotation from the applied load W, as in Figure 9.4.

In Figure 9.4, hydrodynamic flow into recess 4 causes the recess pressure to increase. There is a positive net inflow because the inter-recess film thickness, h_{3-4}, is greater than the inter-recess film thickness, h_{1-4}. Net flow into a recess due to speed may be estimated from

$$q_{4hd} = 1/2\pi DN(L-a)(h_{3-4} - h_{1-4}) \tag{9.9}$$

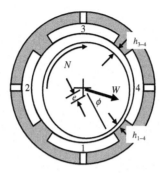

Figure 9.4: Attitude Angle and Inter-Recess Flow Due to Speed.

It is interesting to examine the ratio of hydrodynamic flow into a recess to hydrostatic flow from a recess from equations (9.2) and (9.9). It is found that

$$\frac{q_{hd}}{q_{hs}} = 12\frac{(L-a)}{D}\frac{an}{D}\left(\frac{h_{3-4}}{h_0} - \frac{h_{1-4}}{h_0}\right)S_h \tag{9.10}$$

where

$$S_h = \frac{\eta N}{P_s}\left(\frac{D}{2h_o}\right)^2$$

This shows that increased load due to hydrodynamics depends on S_h. Optimizing so that friction power = pumping power and $K = 1$, the value of S_h is termed S_{ho}. The ratio in equation (9.10) is approximately 35% for a typical recess at 90° to the line of eccentricity, where $L/D = 1$ and $\varepsilon = 0.5$. Film thickness values are given by

$$\frac{h_{3-4}}{h_o} - \frac{h_{1-4}}{h_o} = (1 - \varepsilon\cos\phi_{3-4}) - (1 - \varepsilon\cos\phi_{1-4}) = \varepsilon(\cos\phi_{3-4} - \cos\phi_{1-4}) \tag{9.11}$$

so that $h_{3-4}/h_o - h_{1-4}/h_o = 0.7071$ for $\varepsilon = 0.5$, $\phi_{1-4} = 45°$, and $\phi_{3-4} = 135°$.

It may be seen that net hydrodynamic flow into a recess is maximum when the recess is at 90° to the line of eccentricity. If $\phi_{3-4} = -45°$ and $\phi_{1-4} = +45°$ corresponding to a recess under the line of eccentricity, the net hydrodynamic flow from equation (9.11) is zero.

Inter-recess flow and the effect on bearing pressures together account for the attitude angle between W and ε shown in Figure 9.4.

Flow Data

Concentric flow data are given in Figure 9.5a for land-width ratio $a/L = 0.25$. Flow is slightly larger with a small bridge angle, $\theta = 10°$, compared with flow at the bridge angle, $\theta = 30°$. Concentric flows at $\theta = 10°$ are close to the thin-land values given by equation (9.1).

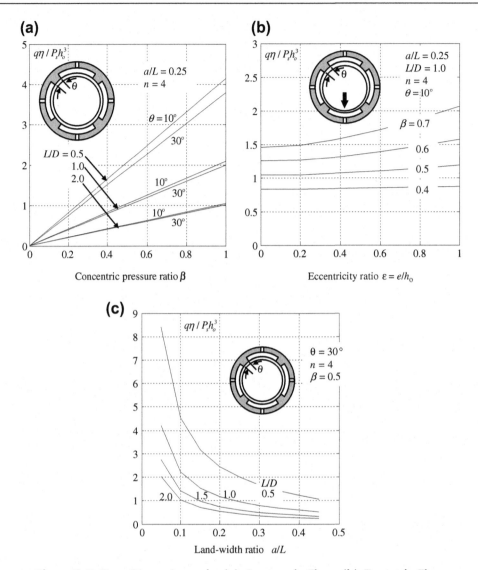

Figure 9.5: Four-Recess Journals: (a) Concentric Flow; (b) Eccentric Flow; (c) Flow vs. Land-Width Ratio a/L.

Flow increases with eccentricity ratio, particularly at high values of pressure ratio (Figure 9.5b). Flow increases greatly at small land widths (Figure 9.5c).

9.3 Load

Approximate Load

The pressure in each recess and the surrounding land area acts approximately on a projected area $(L - a)D\sin(\pi/n)$. The approximate thrust of the ith recess is therefore

$$W_i = P_i(L - a)D \sin(\pi/n) \tag{9.12}$$

where P_i is the pressure in the ith recess. For finite difference solutions, individual pressures are multiplied by elemental projected areas applying the same principle.

Load Resolution

Total load is the sum of the individual pad thrusts, each resolved into the load direction. The load may be resolved in any two orthogonal directions, such as the horizontal and vertical directions:

$$W = \sqrt{W_h^2 + W_v^2} \tag{9.13}$$

From Figure 9.2,

$$W_v = \sum W_i \cos\phi_i \tag{9.14}$$

$$W_h = \sum W_i \sin\phi_i \tag{9.15}$$

where ϕ_i is the angle between the ith recess and the vertical direction.

Load Data and Safe Design Loads

The pressure field may be solved accurately from the Reynolds equation (Chapter 2). Recess pressures are solved simultaneously with pad pressures until sufficient accuracy is achieved. Pad pressures are then resolved and integrated over the surface area. The load is computed and presented in the form:

$$\frac{W}{P_s LD} = \overline{W} \tag{9.16}$$

Computed load data are given in Figure 9.6a–c showing effects of eccentricity ratio and land width.

Safe Maximum Load

The safe maximum load depends on several factors as detailed below. It is generally recommended that a safe maximum load is taken as the value based on an eccentricity ratio of 0.5. This allows for errors in manufacture, in setting pressure ratio and in directionality of loading.

Number of Recesses

Figure 9.6a is for four recesses and Figure 9.6b is for six recesses. Employing six recesses instead of four recesses increases load support at small L/D ratios but the advantage is diminished at $L/D = 2$ by circumferential flow between the recesses.

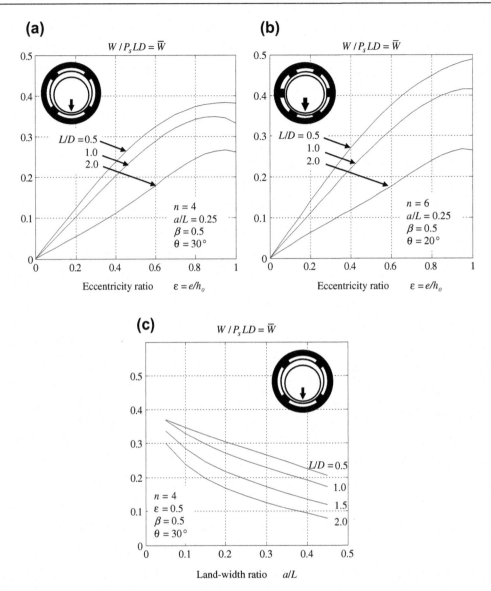

Figure 9.6: (a) Four-Recess Journals, W vs. ε. (b) Six-Recess Journals: W vs. ε. (c) Four-Recess journals: W vs. a/L.

Land-width Ratio

Figure 9.6c shows that smaller land-width ratios give slightly higher load support than the recommended value $a/L = 0.25$. However, flow is greatly increased, as shown in Figure 9.5c.

Computed load data are given in Figure 9.7a–d showing effects of concentric pressure ratio for different eccentricity ratios, directions of loading, and values of power ratio.

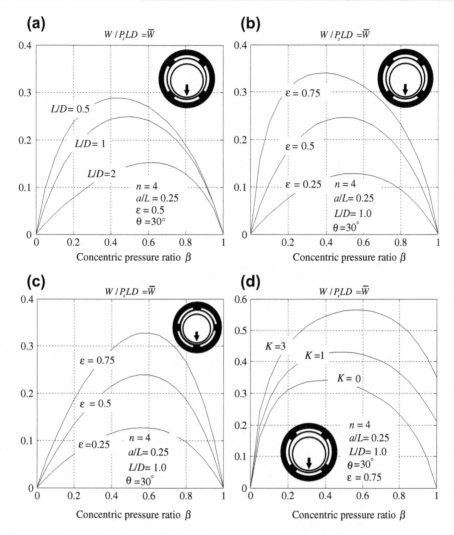

Figure 9.7: (a) Effect of Concentric Pressure Ratio and *L/D* Ratio. (b) Effect of Eccentricity Ratio and Loading Mid-Recess. (c) Effect of Eccentricity Ratio and Loading Mid-Land. (d) Effect of Speed Expressed by Power Ratio *K*.

Concentric Pressure Ratio

Load support varies with concentric pressure ratio, β, as shown in Figure 9.7a–c. Varying pressure ratio within the recommended range, from 0.4 to 0.7, varies load support at $\varepsilon = 0.5$ by a relatively small percentage of less than 10% when loading directly into a recess. The maximum reduction in load support occurs at $\beta = 0.7$ on the curve for $\varepsilon = 0.75$. Load reduction in this case is approximately 20%. A reduced load support when $\beta = 0.7$ is acceptable if maximum load is based on $\varepsilon = 0.5$ rather than on $\varepsilon = 0.75$.

Direction of Loading

Depending on the number of recesses, bearing stiffness and loads may be reduced. The effect of loading direction may be seen by comparing Figure 9.7b and c for four-recess bearings. Maximum load support varies depending on the direction of the line of eccentricity, either towards an inter-recess land or towards a recess center. Loading towards an inter-recess land increases eccentricity at low pressure ratio but reduces it at high pressure ratio. Variations with direction of loading are reduced with five or six recesses instead of four.

Effect of Speed

O'Donoghue et al. (1971) showed that hydrodynamics significantly increase load support at speed. At zero speed, the power ratio $K=0$ but at speed K is increased. The effect of speed is demonstrated in the simplest way by use of power ratio, $K=H_f/H_p$, the ratio of friction power to pumping power described in Section 9.4. When the speed defined by K is in the optimum range $1<K<3$, bearing power is within 15% of the minimum for bearings operating at speed. The collapse load for a bearing at the maximum speed, $K=3$, $a/L=0.25$, and $L/D=1.0$ may be increased by up to 60%, although this effect is not so significant for thin-land recessed bearings.

An important consideration is whether the maximum load must be supported during starting and stopping journal rotation. Recessed bearings are usually designed on the basis of maximum load at zero speed. This allows a degree of additional loading to be applied when operating at speed.

Figure 9.7d shows load support for typical four-recess bearings at speeds corresponding to $K=0$, $K=1$, and $K=3$. The maximum load supported at speed with $K=3$ is some 60% higher than at zero speed $K=0$. Plain hybrid bearings described in Chapter 10 demonstrate even larger increases in maximum load supported at speed.

The following example illustrates use of load data for bearing design.

■ *Example 9.3 Load from Figures 9.6 and 9.7*

Calculate a safe load based on eccentricity ratio $\varepsilon=0.5$ for the two bearings in Example 9.1. The bearings are to be manufactured with four recesses, $n=4$.

Solution

Bearing A: L/D = 0.6 and n = 4

From Figure 9.6a, by interpolation for $L/D=0.6$, $\beta=0.5$, and $a/L=0.25$:

$$\frac{W}{P_sLD}=0.28$$

$$W=2\times10^6\times30\times10^{-3}\times50\times10^{-3}\times0.28=840\text{ N (189 lbf)}$$

However, the bearing specification is for $a/L = 6/30 = 0.2$, whereas Figure 9.6a is for $a/L = 0.25$. The variation of load with a/L could be ignored since the load is only slightly increased if a/L is less than 0.25. This can be checked by noting that the load at an eccentricity ratio up to 0.5 is approximately proportional to $1 - a/L$. A corrected value of W is approximately

$$W_{0.2} = W_{0.25} \times \frac{1 - 0.2}{1 - 0.25} = 1.07 = 840 \times 1.07 = 896 \text{ N } (201 \text{ lbf})$$

The value for $a/L = 0.2$ can also be estimated from Figure 9.6c which shows variations of load with land-width ratio. The dimensionless load is given as approximately 0.3 so that $W = 900$ N (202 lbf).

Bearing B: $L/D = 0.5$, $a/L = 0.25$, and $\beta = 0.4$

From Figure 9.7a, for $\beta = 0.4$:

$$\frac{W}{P_s LD} = 0.29$$

$$W = 300 \times 1 \times 2 \times 0.29 = 174 \text{ lbf}$$

■

9.4 Power, Power Ratio, and Temperature Rise

Power and temperature rise are discussed more fully in Chapter 3. The following is a summary for journal bearings.

Pumping power. Power to pump liquid lubricant through the bearing is

$$H_p = P_s q \tag{9.17}$$

Friction power. Power required to drive the journal and overcome the fluid drag in the bearings is

$$H_f = \frac{\eta A_f U^2}{h_o} \tag{9.18}$$

where, for recessed bearings, $A_f = A - {}^3\!/_4 A_r$, $A = \pi DL$, and $U = \pi DN$.

Minimum power. Total power is

$$H_t = H_p + H_f \tag{9.19}$$

Minimum total power requires that the power ratio lies in the range:

$$1 \le K \le 3 \tag{9.20}$$

where $K = H_f/H_p$.

Temperature rise. Maximum temperature rise as liquid lubricant passes through the bearing is

$$\Delta T = \frac{P_s(1+K)}{Jc\rho} \tag{9.21}$$

where J, the mechanical equivalent of heat, is equal to 1 in a consistent set of units such as SI.

For moderate supply pressure, e.g. $P_s = 3 \text{ MN/m}^2$ (450 lbf/in^2) and optimized bearings, i.e. $1 \leq K \leq 3$, the maximum temperature rise will be typically 6.6 °C. This is insufficient to warrant computations for variable viscosity.

Calculations of flow rate should be based on the viscosity at the estimated average temperature. If the bulk temperature in the tank is T_b:

$$T_{av} = T_b + \frac{1}{2}\Delta T \tag{9.22}$$

9.5 Land-Width Ratios and Concentric Pressure Ratio

Thick-land bearings, $a/L > 0.25$, support a lower hydrostatic load and cause more heat at speed. Thick lands are less easily damaged than thin lands, however. If the land width is $a = 2$ mm in a bearing of diameter 50 mm, a deep scratch would lead to excessive flow rate in that area. For practical reasons, therefore, as well as the efficiency reasons discussed in Chapter 4, the recommended land-width ratio is

$$\frac{a}{L} = 0.25$$

The inter-recess land may be specified by the width b shown in Figure 9.1 or the angle θ in Figure 9.5. It is generally recommended that, for $a/L = 0.25$, the inter-recess angle should be $\theta \approx 30°$ for four-recess bearings or $\theta \approx 20°$ for six-recess bearings. Inter-recess angle or land width b should not be made too small as this increases circumferential flow and detracts from bearing stiffness.

Load considerations based on Figure 9.7 demand that the pressure ratio should not vary too widely from $\beta = 0.5$. Tolerances need to be determined for the clearance between the journal and the bearing. A suitable range for pressure ratio is $0.4 \leq \beta \leq 0.7$. This range may be ensured by selection of suitable tolerances if required, as shown below.

9.6 Selection of Tolerances for Bearing Film Clearance

Maximum and minimum size limits are normally required for product manufacture. In bearing manufacture, it is more helpful to specify limits for the bearing film clearance.

Table 9.1: Suggested tolerances

	Upper Limit	Lower Limit
Clearance	$h_o(U) = 1.5h_o(L)$	$h_o(L)$
Concentric pressure ratio	$\beta = 0.4$	$\beta = 0.7$
Power ratio	$K = 1$	$K = 3$

Stout and Rowe (1974) determined suitable ranges for bearing film clearances. The difference between an upper and lower limit is the tolerance, as explained in Chapter 6. The selection of a suitable tolerance and concentric pressure ratio may be ensured for zero-speed or high-speed bearings by designing for the maximum or minimum clearance given in Table 9.1. For high-speed bearings it is additionally possible to ensure the minimum power condition is achieved.

9.7 Selection of Supply Pressure, Viscosity, and Clearance

These are related variables affecting flow rate, pumping power, and friction power. Selection is aided by use of charts such as Figure 9.8. Starting from the top left-hand corner, in Figure 9.8a a typical set of design decisions are shown leading up to resulting values for clearance limits.

In the first instance, it is decided to use a land-width ratio $a/L = 0.25$. This gives $S_{ho} = 0.05$. Taking a line across to the top right-hand diagram gives upper and lower recommended values of $(P_s/\eta N)(2h_o/D)^2(D/L)$. The upper value corresponds to $K = 1$ and $\beta = 0.4$, whereas the lower value corresponds to $K = 3$ and $\beta = 0.7$.

For a bearing diameter of 50 mm, a line is taken in the lower left-hand diagram, for example to the curves for the IT5 clearance limits. This indicates that the lower clearance ratio should be approximately $2h_o/D = 0.001$ and the upper clearance limit is approximately $2h_o/D = 0.0015$. Multiplying these values by the 50 mm diameter gives clearance limits of 0.05 and 0.075 mm. Dividing by 2, limits of radial clearance are 0.025 and 0.037 mm. In British ips units, these values are 0.001 and 0.0015 in.

Following the lines across to the lower right-hand side, to one of the curves, gives a value . If IT6 clearance values are chosen instead of IT5 values, it can be seen that $(P_s/\eta N)(D/L) = 5 \times 10^6$ may be employed.

Further values of S_{ho} are given in Figure 9.8b for a value of concentric pressure ratio $\beta = 0.5$ and $K = 1$.

Supply Pressure

If the size of the bearing is fixed by machine requirements such as maximum or minimum shaft diameter or length, the supply pressure must be selected so that it is sufficient to

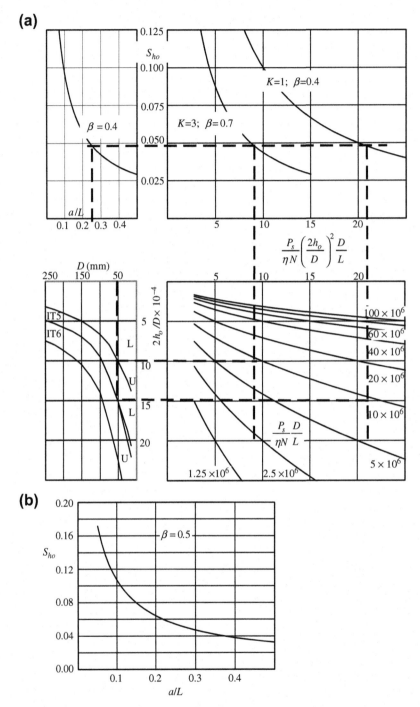

Figure 9.8: (a) Guide to Upper and Lower Clearance Limits. (b) S_{ho} Values for $\beta = 0.5$, $K = 1$.

carry the load. This may be carried out using the data in Figures 9.6 and 9.7. Where the size of the bearing may be varied, the designer will probably select a convenient pressure based on available pumps. Very quiet pumps are available for supply pressures up to 3 MN/m^2 (450 lbf/in^2). Too high a supply pressure increases the temperature rise per pass, although it reduces the bearing diameter and length, and hence reduces the friction power.

Viscosity

Low-speed Bearings (K < 1)

For zero-speed bearings, high viscosity reduces flow and hence pumping power. Viscosity may be chosen for convenience to suit the hydraulic equipment. A suitable value is likely to be in the region of 35 cP at 28 °C.

High-speed Bearings (1 < K < 3)

At high speed, viscosity usually depends on the minimum clearance. The two parameters are related and depend also on supply pressure. Large bearings running at high speed lead to high power consumption and require high flow and low viscosity to achieve cool running. In some cases a designer may consider using air as the fluid medium for very high speeds. Low viscosity requires very small clearances to avoid excessive flow and the associated cost.

Clearance

Clearance has the strongest effect on flow. Increasing clearance by 25% increases flow by 95%, as seen from equation (9.1). For zero-speed bearings the designer will select the smallest clearance compatible with the accuracy achievable in manufacture or deflections, which may occur in service due to load, temperature, and material stability. Errors such as misalignment, ovality, and surface roughness should be small in comparison to the clearance. Ideally such errors should never take up more than 10% of the clearance. One of the attractive features of hydrostatic bearings is their ability to cope with such errors and produce a smoothness of motion that is better than would be expected from the amplitude of the errors.

Use of the Selection Chart (Figure 9.8)

The selection chart allows the designer to determine a suitable value of $P_s/\eta N$ in relation to the desired clearance limits. These are specified on the chart as $U = 2h_o/D$ at the upper limit and $L = 2h_o/D$ at the lower limit. The use of the chart is best illustrated by an example.

■ *Example 9.4 On the Use of the Selection Chart*

It is required to design a four-recess journal bearing for the following:

Maximum load	4 kN (899 lbf)
Rotational speed	20 rev/s
Bearing diameter	64 mm maximum (2.52 in)
Tolerance grade	IT5
Maximum supply pressure	5 MN/m^2 (725 lbf/in^2)
Viscosity	0.010 N s/m^2 minimum (1.45 × 10^6 reyn)

Design parameters may be adjusted within the above constraints.

Solution

From Figure 9.7, $W/P_sLD = 0.25$ at $\varepsilon = 0.5$, $L/D = 1$, and $\beta = 0.4$:

$$P_s = \frac{W}{0.25LD} = \frac{4000}{0.25 \times 0.064 \times 0.064} = 3.9 \text{ MN/m}^2$$

Let $P_s = 4$ MN/m^2 (580 lbf/in^2).

From Figure 9.8a:

1. For $a/L = 0.25$ in the top left-hand part of Figure 9.8 select the appropriate point on the curve for S_{ho}.
2. Proceed horizontally from S_{ho} to intersect the lines for $K = 1$ and $K = 3$ in the top right-hand quadrant.
3. Draw lines vertically downwards from the $K = 1$ and $K = 3$ intersections. The vertical lines will intersect curves of $(P_s/\eta N)(D/L)$.
4. Commencing in the lower left-hand quadrant, take a vertical line from the given diameter (64 mm) to the $2h_o/D$ limits for the desired tolerance grade (IT5). The limits for $2h_o/D$ are 0.0010 and 0.0015. The corresponding clearance values for $2h_o$ are 0.0010 × 64 = 0.064 and 0.096 mm (0.0025 and 0.0038 in). *Note: The selection of clearance and the correct ratio of* h$_o$max/h$_o$min = 1.5 *is critical. It is not important to read the value of* 2h$_o$ *to better than two significant figures, but it is important not to exceed the ratio 1.5:1 for maximum/minimum clearance.*
5. Proceed horizontally from each $2h_o/D$ limit to intersect the previously marked vertical lines in the lower right-hand quadrant. This will provide two intersections on a common $(P_s/\eta N)(D/L)$ curve and hence the viscosity for a given bearing rotational speed and supply pressure. From Figure 9.8, by a process of interpolation, it is found that $(P_s/\eta N)(D/L) \approx 8.5 \times 10^6$. For $N = 20$ rev/s, $P_s = 4$ MN/m^2, $L/D = 1.0$:

$$\frac{4 \times 10^6 \times 1}{\eta \times 20} = 8.5 \times 10^6$$

so that $\eta = 0.0235$ N s/m^2 (3.4 × 10^{-6} reyn).

If viscosity is constrained by the design specification, it may be sufficient to modify a/L to achieve a satisfactory design. An alternative allowing a higher viscosity involves selection of a larger clearance to satisfy the constrained value of $(P_s/\eta N)(D/L)$. Thus, the procedure is to draw lines horizontally from the intersections on the desired $(P_s/\eta N)(D/L)$ curve to determine suitable $2h_o/D$ limits. This alternative procedure leads to increased flow.

Flow

Maximum concentric flow through the bearing may be calculated by reference to Figure 9.5a for the condition $\beta = 0.4$, $2h_o\text{max} = 0.096$ mm, and $P_s = 4$ MN/m^2:

$$h_o = 0.5 \times 0.096 = 0.048 \text{ mm}$$

$$\frac{q\eta}{P_s h_o^3} = 0.84 \quad \text{so that} \quad q = \frac{4 \times 10^6 \times 0.048^3 \times 10^{-9}}{0.0235} \times 0.84$$

$$= 15.8 \times 10^{-6} \text{ m}^3/\text{s} \ (0.95 \ \text{l/min})$$

Power dissipation

Pumping power $H_p = P_s q = 4 \times 10^6 \times 15.8 \times 10^{-6} = 63.2$ watts. When $\beta = 0.4$, $K = 1$, and hence friction power $H_f = KH_p = 63.2$ watts, the total power required $= H_f + H_p = 126.8$ watts.

■

9.8 Bearing Film Stiffness

Bearing film stiffness is the rate at which applied force must be increased or reduced to cause reduction or increase in journal eccentricity. Mathematically, λ (pronounced lambda) is written as the rate of increase of applied load per unit deflection:

$$\lambda = -\frac{dW}{de} \tag{9.23}$$

Stiffness is termed λ_o when the journal is concentric. Analytical expressions may be obtained for concentric stiffness for various forms of flow control. Expressions are obtained applying techniques accurate only for thin-land bearings. Dimensionless stiffness factors are presented in the form:

$$\overline{\lambda} = \frac{\lambda h_o}{P_s LD} \tag{9.24}$$

Concentric Journal Bearing Stiffness (Analytical Method)

A method is described in an appendix to this chapter leading to equations (9.26)–(9.36). The derivation illustrates the action of a journal and also a direct method of obtaining bearing stiffness for various types of flow control.

Analytical results are extended and generalized in the chapter on dynamics and provide a basis for investigation of natural frequency and damping.

The analytical method complements the more exact method of numerical solution of bearing pressures and differentiation of the load/eccentricity results.

The analytical method described in the appendix to this chapter (Section 9.10) leads to

$$\bar{\lambda}_o = \frac{\lambda_o h_o}{P_s LD} = \frac{4.30(1 - a/L)\beta(1 - \beta)}{1 + 0.5\gamma(1 - \beta)} \tag{9.25}$$

where $\gamma = na(L - a)/\pi Db$ (pronounced *gamma*) is a circumferential flow factor.

In general, the derivation may be simplified after working out the dimensionless stiffness due to one recess because it may be shown that

$$\bar{\lambda}_o = \frac{n}{2}\bar{\lambda}_1 \tag{9.26}$$

Rowe (1980) obtained a general equation for hydrostatic stiffness with any number of recesses and with capillary, orifice, constant flow, or diaphragm control:

$$\lambda = \frac{P_s LD}{h_o} \frac{3n^2}{2\pi} \frac{\beta(1 - a/L)\sin^2(\pi/n)}{z + 1 + 2\gamma\sin^2(\pi/n)} \tag{9.27}$$

where

Capillary : $\quad z = \dfrac{\beta}{1 - \beta}$

Orifice : $\quad z = \dfrac{1}{2}\dfrac{\beta}{1 - \beta}$

Constant flow : $\quad z = 0$

Diaphragm control : $\quad z = \dfrac{\beta}{1 - \beta} - \dfrac{6\beta}{\bar{\lambda}_d}$

Expressions for hydrodynamic stiffness and squeeze film damping are given in Chapter 14 on dynamics.

Effect of Flow Control on Stiffness

Concentric stiffness factors for bearings with and without axial dividing slots are given in Tables 9.2 and 9.3. It may be seen that stiffness increases with the number of recesses, although this is found to be a law of diminishing returns. The factor K_{bs} takes account of the bearing shape with respect to land-width ratio for nongrooved journal bearings and also with respect to angular extent of the pads for grooved journals.

Table 9.2: Concentric stiffness $\bar\lambda_o$ for nongrooved journals (without axial slots)

n	Capillary	Orifice	Constant Flow	Diaphragm Valve
3	$\dfrac{3.22K_{bs}\beta(1-\beta)}{1+1.5\gamma(1-\beta)}$	$\dfrac{6.44K_{bs}\beta(1-\beta)}{2-\beta+3\gamma(1-\beta)}$	$\dfrac{3.22K_{bs}\beta}{1+1.5\gamma}$	$\dfrac{3.22K_{bs}\beta(1-\beta)}{1+1.5\gamma(1-\beta)-\dfrac{6\beta(1-\beta)}{\bar\lambda_d}}$
4	$\dfrac{3.82K_{bs}\beta(1-\beta)}{1+\gamma(1-\beta)}$	$\dfrac{7.65K_{bs}\beta(1-\beta)}{2-\beta+2\gamma(1-\beta)}$	$\dfrac{3.82K_{bs}\beta}{1+\gamma}$	$\dfrac{3.82K_{bs}\beta(1-\beta)}{1+\gamma(1-\beta)-\dfrac{6\beta(1-\beta)}{\bar\lambda_d}}$
5	$\dfrac{4.12K_{bs}\beta(1-\beta)}{1+0.69\gamma(1-\beta)}$	$\dfrac{8.25K_{bs}\beta(1-\beta)}{2-\beta+\gamma(1-\beta)}$	$\dfrac{4.12K_{bs}\beta}{1+0.69\gamma}$	$\dfrac{4.12K_{bs}\beta(1-\beta)}{1+0.69\gamma(1-\beta)-\dfrac{6\beta(1-\beta)}{\bar\lambda_d}}$
6	$\dfrac{4.30K_{bs}\beta(1-\beta)}{1+0.5\gamma(1-\beta)}$	$\dfrac{8.60K_{bs}\beta(1-\beta)}{2-\beta+\gamma(1-\beta)}$	$\dfrac{4.30K_{bs}\beta}{1+0.5\gamma}$	$\dfrac{4.30K_{bs}\beta(1-\beta)}{1+0.5\gamma(1-\beta)-\dfrac{6\beta(1-\beta)}{\bar\lambda_d}}$

where $K_{bs}=(1-a/L)$ for nongrooved journals

Table 9.3: Dimensionless stiffness $\bar\lambda_o$ for grooved journals (with axial slots)

n	Capillary	Orifice	Constant Flow	Diaphragm Valve
3	$\dfrac{4.5K_{bs}\beta(1-\beta)}{1+\gamma(1-\beta)}$	$\dfrac{9K_{bs}\beta(1-\beta)}{2-\beta+2\gamma(1-\beta)}$	$\dfrac{4.5K_{bs}\beta}{1+\gamma}$	$\dfrac{4.5K_{bs}\beta(1-\beta)}{1+\gamma(1-\beta)-\dfrac{6\beta(1-\beta)}{\bar\lambda_d}}$
4	$\dfrac{6K_{bs}\beta(1-\beta)}{1+\gamma(1-\beta)}$	$\dfrac{12K_{bs}\beta(1-\beta)}{2-\beta+2\gamma(1-\beta)}$	$\dfrac{6K_{bs}\beta}{1+\gamma}$	$\dfrac{6K_{bs}\beta(1-\beta)}{1+\gamma(1-\beta)-\dfrac{6\beta(1-\beta)}{\bar\lambda_d}}$
5	$\dfrac{7.5K_{bs}\beta(1-\beta)}{1+\gamma(1-\beta)}$	$\dfrac{15K_{bs}\beta(1-\beta)}{2-\beta+2\gamma(1-\beta)}$	$\dfrac{7.5K_{bs}\beta}{1+\gamma}$	$\dfrac{7.5K_{bs}\beta(1-\beta)}{1+\gamma(1-\beta)-\dfrac{6\beta(1-\beta)}{\bar\lambda_d}}$
6	$\dfrac{9K_{bs}\beta(1-\beta)}{1+\gamma(1-\beta)}$	$\dfrac{18K_{bs}\beta(1-\beta)}{2-\beta+2\gamma(1-\beta)}$	$\dfrac{9K_{bs}\beta}{1+\gamma}$	$\dfrac{8K_{bs}\beta(1-\beta)}{1+\gamma(1-\beta)-\dfrac{6\beta(1-\beta)}{\bar\lambda_d}}$

where $K_{bs}=\sin\theta\left[\dfrac{\sin\theta}{\theta}+\gamma\cos\theta\right](1-a/L)$ for grooved bearings (see Figure 9.1)

Concentric stiffness is least with capillary flow control and is increased when employing orifice control and further increased with constant flow. A range of stiffness values can be obtained with diaphragm control ranging from capillary values through to infinite stiffness, as described in Chapter 5.

Stiffness varies with direction of loading, particularly for four-recess bearings. Stiffness values are given for loading towards a recess. If the shaft is moved towards the center of an inter-recess land, the stiffness may be lower depending on the pressure ratio.

For a four-recess bearing with a pressure ratio of 0.5 a typical reduction would be about 5–10%, but with low values of β and high values of γ the reduction may be as high as 35%. Rowe (1980) found that increasing the number of recesses to five or six reduces the directionality effect dramatically to less than 2%.

Computed Stiffness Data

Values of stiffness computed by numerical methods are given in Figure 9.9a and b. The figures show that stiffness follows trends similar to those of opposed pads described in Chapter 7.

Maximum concentric stiffness is found when β is approximately 0.55. This is in the middle of the range recommended when making allowance for a clearance tolerance, i.e. $0.4 < \beta < 0.7$.

Stiffness may either increase or reduce with increasing eccentricity depending on concentric pressure ratio. Stiffness increases with eccentricity with very low values of pressure ratio. Stiffness reduces with eccentricity for high values of pressure ratio.

The average dimensionless stiffness across the range of eccentricities is, by definition, equal to the maximum dimensionless load support at $\varepsilon = 1$. With $\beta = 0.5$, dimensionless stiffness typically varies between 0.25 and 0.35.

At maximum eccentricity, stiffness drops very significantly and in some cases even becomes negative. Negative stiffness indicates collapse of bearing load support. It is therefore recommended that maximum load should be based on an eccentricity ratio not exceeding the value $\varepsilon = 0.5$.

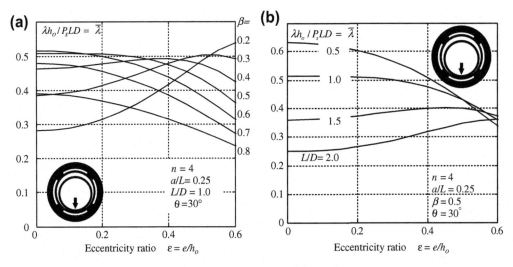

Figure 9.9: (a) Capillary-Controlled Journals: Stiffness for Various Pressure Ratios.
(b) Capillary-Controlled Journals: Stiffness for Various L/D Ratios.

■ Examples 9.5a and 9.5b Bearing Film Stiffness

Example 9.5a British ips units and analytical formulae
To calculate concentric bearing film stiffness for capillary control and also for orifice control for the following data:

$n = 6$ recesses
$D = 3$ in shaft diameter
$a = 0.2$ in axial flow land
$b = 0.3$ in inter-recess land
$L = 3$ in bearing length
$P_s = 150$ lbf/in^2 supply pressure
$p_r = 75$ lbf/in^2 unloaded condition recess pressure
$2h_o = 0.0024$ in diametral clearance

Solution
From equation (9.27) and Table 9.2:

$$\gamma = \frac{na(L-a)}{\pi Db} = \frac{6 \times 0.2 \times (3-0.2)}{\pi \times 3 \times 0.3} = 1.2$$

$$\bar{\lambda}_o = \frac{4.30 \times (1-0.067) \times 0.5(1-0.5)}{1 + 0.5 \times 1.2 \times (1-0.5)} = 0.77$$

$$\lambda = \frac{P_s LD}{h_o}\bar{\lambda}_o = \frac{150 \times 3 \times 3}{0.0012} \times 0.77 = 870,000 \text{ lbf/in}$$

For orifice control:

$$\bar{\lambda}_o = \frac{8.60 \times (1-0.067) \times 0.5 \times (1-0.5)}{2 - 0.5 + 1.2 \times (1-0.5)} = 0.955$$

$$\lambda = \frac{P_s LD}{h_o}\bar{\lambda}_o = \frac{150 \times 3 \times 3}{0.0012} \times 0.955 = 1.08 \times 10^6 \text{ lbf/in}^2$$

Example 9.5b SI units and computed data including clearance limits
Bearing film stiffness at eccentricity ratio $\varepsilon = 0.5$ with capillary control for:

$n = 4$ recesses
$D = 64$ mm shaft diameter
$a = 16$ mm axial flow land
$\theta = 30°$ inter-recess land
$L = 64$ mm bearing length
$P_s = 4$ MN/m^2 supply pressure
$2h_o = 0.080$ mm diametral clearance, $\beta = 0.4$
$2h_o = 0.054$ mm diametral clearance, $\beta = 0.7$

Solution

At $\beta = 0.4, \bar{\lambda} = 0.475$:

$$\lambda = \frac{P_s LD}{h_o}\bar{\lambda} = \frac{4 \times 10^6 \times 64 \times 10^{-3} \times 64 \times 10^{-3}}{0.04 \times 10^{-3}} \times 0.475 = 195 \text{ MN/m}$$

At $\beta = 0.7, \bar{\lambda} = 0.33$:

$$\lambda = \frac{P_s LD}{h_o}\bar{\lambda} = \frac{4 \times 10^6 \times 64 \times 10^{-3} \times 64 \times 10^{-3}}{0.04 \times 10^{-3}} \times 0.33 = 135 \text{ MN/m}$$

■

References

O'Donoghue, J. P., & Rowe, W. B. (1968). Hydrostatic journal bearing (exact procedure). *Tribology International, 1*(4; Nov), 230–236.

O'Donoghue, J. P., Rowe, W. B., & Hooke, C. J. (1970a). Computer analysis of externally pressurized journal bearings. 1969 Tribology Convention, Brighton. *Proceedings of the Institution of Mechanical Engineers, 184*(Part 3L), 48–53.

O'Donoghue, J. P., Rowe, W. B., & Hooke, C. J. (1970b). Some tolerancing effects in hydrostatic bearings. *Proceedings of the 11th International Machine Tool Design and Research Conference,* September. Oxford: Pergamon Press.

O'Donoghue, J. P., Hooke, C. J., & Rowe, W. B. (1971). A solution using the superposition technique for externally pressurized multi-recess journal bearings including hydrodynamic effects. *Proceedings of the Institution of Mechanical Engineers, 185*(5), 57–61.

Rowe, W. B. (1980). Dynamic and static properties of recessed hydrostatic journal bearings by small displacement analysis. *Journal of Lubrication Technology, 102*(1; Jan.), 71–79.

Stout, K. J., & Rowe, W. B. (1974). Externally pressurized bearings − Part 3 Design of hydrostatic bearings including tolerancing procedures. *Tribology International, 7*(5; Oct.), 195–212.

Appendix 1: Derivation of Journal Bearing Stiffness by an Approximate Analytical Method

The derivation of concentric stiffness for any type of journal bearing with any type of control device is similar to the following example for a six-pad capillary-controlled bearing. Each recess, associated axial flow land, and inter-recess land is numbered as in Figure 9.4. Equating flow through the bearing for recess 1 to flow through the restrictor:

$$(P_s - P_1)\frac{\pi d_c^4}{128\eta l_c} = \frac{\overline{B}}{n}\frac{P_1 h_1^3}{\eta} + \frac{\gamma}{2}\frac{\overline{B}}{n}(P_1 - P_2)\frac{h_{1-2}^3}{\eta} + \frac{\gamma}{2}\frac{\overline{B}}{n}(P_1 - P_6)\frac{h_{1-6}^3}{\eta} \qquad (9.28)$$

where allowance for circumferential inter-recess flow is introduced by γ, the ratio of axial flow resistance to circumferential flow resistance in the unloaded condition. For a bearing without axial slots, as in Figure 9.1,

$$\gamma = \frac{na(L - a)}{\pi Db} \qquad (9.29)$$

Equation (9.26) can be simplified by evaluating the capillary factor by reference to the concentric condition when all film thickness values are equal to h_o and all recess pressures are equal to βP_s so that

$$\frac{\pi d_c^4}{128 l_c} = \frac{\beta}{1 - \beta} \frac{\overline{B} h_o^3}{n} \tag{9.30}$$

This value may be inserted in equation (9.28), which may be further simplified for the case where $P_2 = P_6$ and where $h_{1-6} = h_{1-2}$:

$$\frac{\beta}{1 - \beta}(P_s - P_1) = \frac{h_1^3}{h_o^3} P_1 + \gamma \frac{h_{1-2}^3}{h_o^3}(P_1 - P_2) \tag{9.31}$$

If the journal is displaced a small distance in the direction of loading, there will be resulting changes in gap, Δh_n, and small changes in recess pressure, ΔP. The change in gap is not constant around the journal and the values h_i may be expressed as a proportion of the eccentricity e. Since flow is proportional to gap by the third power,

$$(h_o - \Delta h_i)^3 = \frac{1}{\theta_2 - \theta_1} \int_{\theta_1}^{\theta_2} (h_o - e\cos\theta)^3 d\theta \tag{9.32}$$

For $e \ll 1$, i.e. small perturbations,

$$3h_o^2 \Delta h_1 = \frac{3h_o^2}{\theta_2 - \theta_1} e(\sin\theta_2 - \sin\theta_1) \tag{9.33}$$

Thus, for $n = 6$,

$$\Delta h_1 = \frac{3e}{\pi} \tag{9.34}$$

For small perturbations, equation (9.31) may be written as

$$\frac{\beta}{1 - \beta}(P_s - P_o - \Delta P_1) = \frac{(h_o + \Delta h_1)^3}{h_o^3}(P_o + \Delta P_1)$$
$$+ \frac{\gamma(h_o + \Delta h_{1-2})^3}{h_o^3}(P_o + \Delta P_1 - P_o - \Delta P_2) \tag{9.35}$$

Also, for small perturbations it may be assumed that

$$\Delta P_2 = \frac{\Delta h_2}{\Delta h_1} \Delta P_1 \approx 0.5 \quad \text{for } n = 6 \tag{9.36}$$

Evaluating equations (9.31), (9.33), and (9.34) leads to

$$\frac{h_o \Delta P_1}{e} = \frac{2.865 P_s \beta (1 - \beta)}{1 + 0.5\gamma(1 - \beta)} \tag{9.37}$$

Treating other recesses similarly and summing pressure changes using equations (9.12)–(9.14) yields

$$\overline{\lambda}_o = \frac{\lambda_o h_o}{P_s L D} = \frac{4.30(1 - a/L)\beta(1 - \beta)}{1 + 0.5\gamma(1 - \beta)} \tag{9.38}$$

Appendix 2: Tabular Design Procedure

Procedure 9.A1 Hydrostatic Journal Bearings

Step	Symbol	Description of Operation	Notes
1	W	Specify: extreme load	From design requirements
2	P_s	Specify: supply pressure	Suggestion: 2 MN/m^2 (291 lbf/in^2)
3	n	Specify: number of recesses/slots	Suggestion: 4–6 recesses or 12 slots
4	β	Specify: pressure ratio, P_{ro}/P_s	Suggestion: 0.5 or see Chapter 9
5	L/D	Specify: length/diameter	Suggestion: 1 or see Chapter 9
6	a/L	Specify: land-width ratio	Suggestion: 0.25 or see Chapter 9
7	θ	Specify: bridge angle	Suggestion: 30° for $n=4$ or 5° for $n=12$
8	$\overline{W}$	Read: load factor	See Figure 9.6 (or Figure 10.4 for slot entry)
9	$\overline{B}$	Calculate: flow factor	From $(\pi D/L)/(6a/L)$
10	D	Calculate: from $\sqrt{\dfrac{W}{P_s\overline{W}L/D}}$	If diameter D is too large, increase P_s
11	L	Calculate: length	From $D \times L/D$
12	a	Calculate: land width	From land-width ratio
13	A_r	Calculate: total recess area	From $\pi DL(1-2a/L)(1-n\theta°/360)$
14	A_f	Calculate: friction area	From $\pi DL - {}^3/_4 A_f$
15	K	Specify: power ratio	Suggest $K=1$
16	h_o	Specify: radial bearing gap	h_o should be at least 5–10 times flatness tolerance
17	U	Calculate: sliding speed	From πDN
18	η	Calculate: viscosity	From $\eta = \dfrac{P_s h_o^2}{U}\sqrt{\dfrac{K\beta\overline{B}}{A_f}}$
19	q_o	Calculate: concentric flow	$q_o = \dfrac{P_s h_o^3 \beta\overline{B}}{\eta}$
20	$\overline{\lambda}$	Read: stiffness factor	From Figures 9.9 or from Tables 9.2 and 9.3
21	λ	Calculate: stiffness	From $P_s LD\overline{\lambda}/h_o$
22	H_p	Calculate: pumping power	From $H_p = P_s q$
23	H_f	Calculate: friction power	From $H_f = \eta A_f U^2/h_o$
24	ΔT	Calculate: temperature rise/pass	From $0.6 \times 10^{-6}(1+K)P_s$

Example 9.A1 Hydrostatic Journal Bearing to Carry 8900 N at 1000 rev/min with Capillary Control

Step	Symbol	Example of Working	Result
1	W	Extreme load	8900 N (2000 lbf)
2	P_s	Supply pressure	$P_s = 2$ MN/m^2
3	n	Number of recesses	4
4	β	Pressure ratio	0.5
5	L/D	Length/diameter	1.0
6	a/L	Land-width ratio	0.25
7	θ	Bridge angle	30°
8	$\overline{W}$	Load factor from Figure 9.6a	0.25
9	$\overline{B}$	$(\pi \times 1)/(6 \times 0.25)$	2.094
10	D	$\sqrt{\dfrac{8900}{2 \times 10^6 \times 0.25 \times 1}}$	0.1334 m (5.253 in)
11	L	0.1334×1	0.1334 m (5.253 in)
12	a	0.1334×0.25	0.03335 m (1.313 in)
13	A_r	$\pi \times 0.1334^2 \times (1 - 0.5) \times (1 - 4 \times 30/360)$	0.01864 m^2 (28.89 in)
14	A_f	$\pi \times 0.1334^2 - 0.75 \times 0.01864$	0.04193 m^2 (64.99 in^2)
15	K	Power ratio	1
16	h_o	Radial bearing gap	50 μm (0.002 in)
17	U	$\pi \times 0.1334 \times 1000/60$	6.985 m/s
18	η	$\dfrac{2 \times 10^6 \times 50^2 \times 10^{-12}}{6.985}\sqrt{\dfrac{0.5 \times 2.094}{0.04193}}$	0.003577 N s/m^2 (3.577 cP)
19	q_o	$\dfrac{2 \times 10^6 \times 50^3 \times 10^{-18} \times 0.5 \times 2.094}{0.003577}$	73.18×10^{-6} m^3/s (4.39 l/min)
20	$\overline{\lambda}$	Stiffness factor	0.5
21	λ	$\dfrac{2 \times 10^6 \times 0.1334^2 \times 0.5}{50 \times 10^{-6}}$	356 MN/m (2.032×10^6 lbf/in)
22	H_p	$2 \times 10^6 \times 73.18 \times 10^{-6}$	146 W
23	H_f	$0.003577 \times 0.04193 \times 6.985^2/50 \times 10^{-6}$	146 W
24	ΔT	$0.6 \times 10^{-6}(1 + 1) \times 2 \times 10^{-6}$	2.4 °C

CHAPTER 10

Plain Journal Bearings

Summary of Key Design Formulae

$$W = P_s LD \cdot \overline{W}$$ Bearing film load support

$$\lambda = \frac{P_s LD}{h_o} \cdot \overline{\lambda}$$ Bearing film stiffness

$$q_o = \frac{P_s h_o^3}{\eta} \cdot \beta \overline{B}$$ Hydrostatic flow

$$q_o = \frac{P_s h_o^3}{\eta} \cdot \beta \overline{B} \cdot \left(\frac{p_{ro} + p_a}{2p_a} \right)$$ Aerostatic flow

$$K = \frac{H_f}{H_p}$$ Power ratio

Part A Hydrostatic/Hybrid Plain Journal Bearings

10.1 Introduction

Plain journal bearings, illustrated in Figures 10.1 and 10.2, are suitable for hydrostatic, hybrid, and aerostatic designs. Plain nonrecessed bearings are preferred for high speeds and high loads. Plain bearings work well at low and high speeds. The advantage is as follows.

Plain hybrid bearings gain substantial hydrodynamic load support. Unlike a purely hydrodynamic bearing, a plain hybrid bearing supports load at zero speed. This means wear can be avoided on starting and stopping. Plain hybrid bearings are particularly suited where operating load increases with speed. For example, a shaft with a large out-of-balance mass experiences a rotating force $mr(2\pi N)^2$ that increases with speed. Plain hybrid bearings support load roughly increasing in proportion to N. Plain hybrid bearings can also withstand heavy dynamic loads that vary widely in the direction of application and can tolerate loads over and above the zero-speed load.

A great advantage of a plain hybrid bearing compared with the usual recessed bearing is the possibility to design for a smaller shaft diameter, thus reducing initial cost and power consumption.

Hydrostatic, Aerostatic and Hybrid Bearing Design.
DOI: 10.1016/B978-0-12-396994-1.00010-3
207

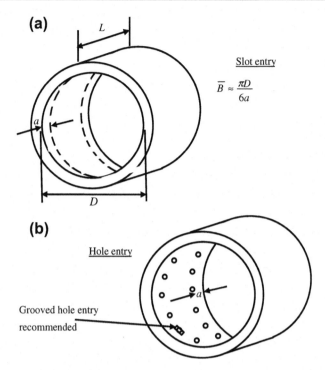

Figure 10.1: Plain Journal Bearings with Double Entry: (a) Slot Entry; (b) Hole Entry.

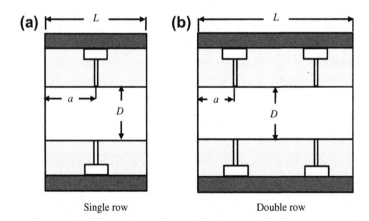

Figure 10.2: Plain Journal Bearings: (a) Single-Row Geometry; (b) Double-Row Geometry.

10.2 Selection of Bearing Configuration

Recesses detract from hydrodynamic support and therefore plain hybrid bearings support higher loads at speed. A further consideration is manufacturing simplicity and cost. The need to machine or fabricate recesses may be simplified or possibly avoided altogether.

At zero speed, hydrostatic support in plain bearings is better with a greater number of entry ports, normally 12. At high speeds and high power ratio, eight entry ports are sufficient.

Slot-Entry Bearings (Shires and Dee, 1967; Rowe and Koshal, 1980; Ives and Rowe, 1987; Rowe et al., 1976, 1977)

Slot-entry restrictors may be formed by a slotted shim assembled into the bearing (see Chapter 1). Relevant dimensions of the entry ports are illustrated in Figures 1.5 and 1.14. Slots may alternatively be machined or etched into or produced by plating the end rings. Each slot is typically separated circumferentially by 5° from adjacent slots.

Hole-Entry Bearings (Rowe et al., 1982; Yoshimoto et al., 1988)

Rows of holes may be employed where each hole feeds from a restrictor. Each hole feeds directly into the bearing or into a narrow slot to spread the pressure. Entry ports are sometimes incorporated into a bearing by assembly of grooved or slotted end rings. However, entry slots must not be allowed to link up circumferentially, which would equalize entry pressures around the bearing.

A bearing with 12 holes per row feeds lands that extend around a total arc of 30°. A 12-hole bearing having 25° entry grooves is identical to a slot-entry bearing having entry slots separated by a 5° arc. Entry grooves for a 12-hole per row bearing may be conveniently designed to extend 10° or 20° to maximize support.

A bearing with eight holes per row has 45° lands. Entry grooves should extend at least 20°. Smaller entry ports are not recommended because too much zero-speed support is sacrificed. Data are provided for both eight- and 12-port arrangements. A larger entry arc is better for zero-speed support as revealed by the data charts.

The selection of hole-entry or slot-entry bearings depends mainly on preference for manufacture and maintenance. Slot-entry bearings need finer filtration to prevent gradual silting of the slot restrictors. Hole-entry bearings are more prone to partial or complete blockage of a restrictor due to a larger single particle in the system. Grooved ports can be employed with hole restrictors and avoid silting, although grooved bearings may also suffer blockages. Ideally, restrictors are removable for maintenance. Adequate filtration is very important.

Slot-entry and hole-entry bearings can be designed as single row or double row. The difference is illustrated in Figure 10.2. Double-row bearings carry higher loads than single-row bearings. To some extent, there is a penalty to be paid for the increased load capacity. Flow is higher with double-row bearings, although there is greater recirculation around the central section. To reduce flow, double-row bearings employ a higher viscosity lubricant or a slightly smaller clearance.

At speed, maximum hybrid load for minimum power is achieved as land width is reduced. A minimum practicable value of a/L is assumed to be 0.1 and this value is used for most high-speed designs.

Asymmetric Hybrid Bearings

For loads applied predominantly in one direction, it is possible to support greater load with less cavitation using an asymmetric design. This desirable outcome can be achieved with a bearing that has high-pressure inlet slots under the load and a low-pressure axial groove opposite (Ives and Rowe, 1987).

10.3 Power Ratio, K

Load support depends very strongly on power ratio K, defined as the ratio of friction power to pumping power, $K = H_f/H_p$. Plain hybrid bearings are very efficient on a load/power basis when designed for power ratio in the following range (Rowe and Koshal, 1980):

$$3 \leq K \leq 9 \qquad \text{recommended for hybrid operation} \qquad (10.1)$$

This range, together with the recommendations below for pressure ratio and clearance, provides for a film thickness tolerance.

For low speeds and purely hydrostatic support, the recommended range is

$$1 \leq K \leq 3 \qquad \text{recommended for hydrostatic operation} \qquad (10.2)$$

10.4 Concentric Hydrostatic Pressure Ratio, β

Concentric pressure ratio is recommended to lie within the range:

$$0.4 \leq \beta \leq 0.7 \qquad (10.3)$$

This range ensures zero-speed load support near the maximum but also means hybrid support is approximately optimal.

When considering a film thickness tolerance, the minimum value of pressure ratio is set to $\beta = 0.4$ at the upper clearance $h_o(U)$ and lower power ratio $K = 3$. Maximum pressure ratio $\beta = 0.7$ then corresponds to the lower clearance $h_o(L)$ and upper power ratio $K = 9$. This is further described below.

10.5 Bearing Clearance and Clearance Limits

Upper and lower limits may be specified for concentric film thickness. These are termed $h_o(U)$ and $h_o(L)$ respectively. A suitable range for hybrid bearings is given in Table 10.1. A chart relating clearance limits to operating conditions is given in Figure 10.3.

Table 10.1: Recommended limits for clearance, pressure ratio, and power ratio

	Upper Limit	Lower Limit
Concentric film thickness	$h_o(U) = 1.5h_o(L)$	$h_o(L)$
Concentric pressure ratio	$\beta = 0.4$	$\beta = 0.7$
Power ratio	$K = 3$	$K = 9$

Figure 10.3 suggests suitable clearance tolerances. Tighter tolerances can be specified if desired to maintain pressure ratio close to $\beta = 0.5$ although, in practice, tolerances are not always held as tight as envisaged at the design stage.

Figure 10.3a and b helps in selecting combinations of supply pressure, viscosity, rotational speed, and bearing clearance to achieve a required value of power ratio K.

Clearance limits may be selected for plain hybrid bearings from Figure 10.3a and b. The procedure is similar to Example 9.4 for recessed bearings except that $K = 3$ and $K = 9$ are employed in Figure 10.3 to obtain a value of $P_s/(\eta N)$. An example is indicated on the chart by the chain line for the land-width ratio $a/L = 0.1$. Comparison with Chapter 9 shows that plain hybrid bearings employ slightly smaller clearances than recessed bearings. The chart provides suitable values of $P_s/(\eta N)$ for particular sizes of bearing.

The same chart can be employed, if so desired, to design plain bearings for the low power ratio range $1 < K < 3$. The same clearance ratio may be employed as in the hybrid design but $P_s/(\eta N)$ is found by drawing a vertical line down from the curve for $K = 1$, instead of from $K = 3$, to meet the horizontal line drawn across from the upper clearance ratio limit (U).

Two sets of tolerance grades are given for the clearance ratio. These roughly correspond to International Tolerance Grades IT5 and IT6.

Larger or smaller clearance ratios may be employed but the proportion 1.5:1 should not be exceeded. This facilitates values of $P_s/(\eta N)$ larger or smaller than indicated on the chart. Adjustment of clearance ratios may ease manufacture or reduce flow as required.

Single-row bearings are designed from Figure 10.3 by selecting $a/L = 0.5$. Single-row bearings employ slightly larger values of bearing film clearance than double-row bearings.

Values of the speed parameter S_h vary with pressure ratio and power ratio. For plain bearings, values are taken from Figure 10.3 or found from

$$S_h = \frac{1}{4\pi L/D} \sqrt{\frac{K\beta}{6a/L}}$$

(a)

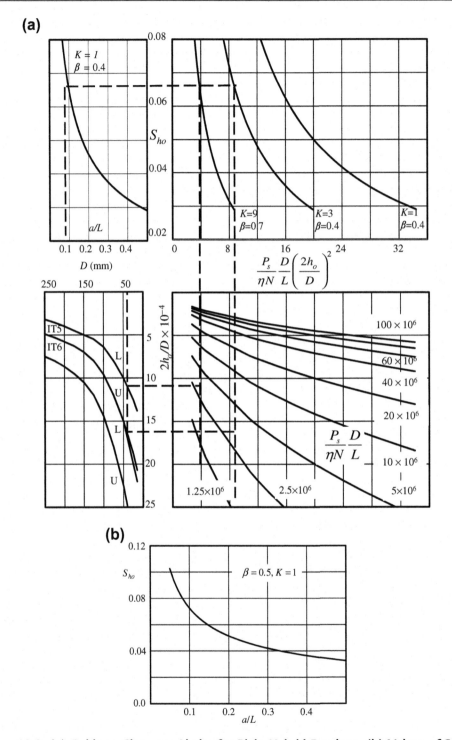

Figure 10.3: (a) Guide to Clearance Limits for Plain Hybrid Bearings. (b) Values of Speed Parameter S_{ho} for $\beta = 0.5$ and $K = 1$.

10.6 Hydrostatic Load Support

Zero-Speed Slot-Entry Bearings with a/L = 0.25 and a/L = 0.5

Load support at zero speed is lower than at high speeds. Design for zero speed implies $K = 0$ as in Chapter 9. Plain bearings and recessed bearings support similar loads at zero speed, as shown by Figures 10.4 and 10.5 compared with Figure 9.7b and c. Slot-entry bearings with 12 slots per row support marginally higher loads.

Figure 10.4 gives load data for slot-entry bearings at zero speed having 12 slots per row. Single-row bearings have $a/L = 0.5$ and double-row bearings have $a/L = 0.25$.

Figure 10.5 gives data for slot-entry bearings having eight slots per row. Load support at zero speed is reduced. However, eight slots per row are adequate for many purposes,

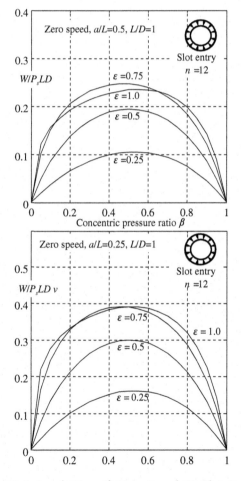

Figure 10.4: Load *W* vs. *β* Zero-Speed 12-Slot Bearings.

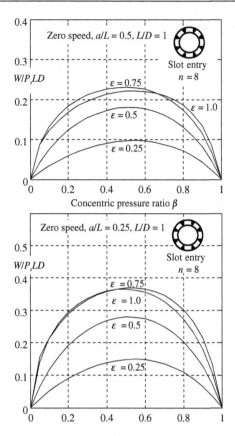

Figure 10.5: Load *W* vs. *β*: Zero-Speed Eight-Slot Bearings.

particularly when the bearing is to be designed for hybrid load support at higher power ratios.

A reasonable guide to zero-speed load support for plain bearings is given by the charts for recessed bearings in Chapter 9, although additional charts for plain bearings are given below.

In Figures 10.4 and 10.5, loads supported at an eccentricity ratio of 1 tend to be lower than or much the same as at an eccentricity ratio of 0.75. Unfortunately, a characteristic of zero-speed bearings, plain and recessed, is that bearing stiffness diminishes to zero approaching touch-down. Maximum load for zero speed is therefore based on an eccentricity ratio of 0.5.

Zero-Speed Slot-Entry and Hole-Entry Bearings with a/L = 0.1

Eight Entry Ports Per Row

Figure 10.6 compares slot-entry and hole-entry bearings for zero-speed load support when $n = 8$ and $a/L = 0.1$. For hybrid load support, a thin-land geometry is recommended,

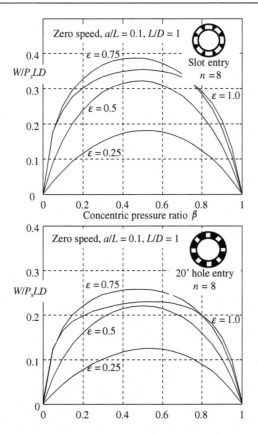

Figure 10.6: Zero-Speed, Eight-Slot, and Eight-Hole Bearings with 20° Ports, at $a/L = 0.1$.

$a/L = 0.1$. Zero-speed charts are presented for hybrid bearings because high-speed bearings must carry sufficient, albeit smaller, load at zero speed. Zero-speed loads supported by slot-entry bearings are superior to loads supported by hole-entry bearings, although a hole-entry bearing with $L/D = 1$ supports more than $W/P_sLD = 0.2$ even at $\beta = 0.7$. For better zero-speed load support, the 12 slots per row option is preferred.

10.7 Hybrid Slot-Entry Bearings

Load Support for Minimum Speed Specified by K = 3

Load support depends on the minimum value of K, since load support at $K = 3$ is lower than at $K = 9$ (Figure 10.7). Therefore, maximum load must be defined at $K = 3$ rather than at $K = 9$.

Effect of Increasing K

A thin-land bearing $a/L = 0.1$ operating at the speed represented by $K = 3$ supports five times more load than the zero-speed bearing given in Figure 10.6. High eccentricity

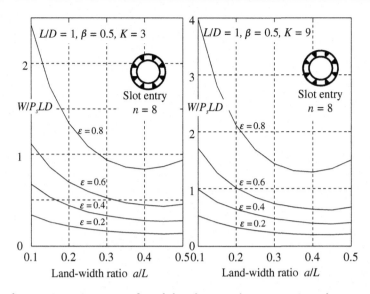

Figure 10.7: *W* vs. *a*/*L* for Eight-Slot Bearings: *K* = 3 and *K* = 9.

ratios may be employed more safely than with zero-speed bearings, where stiffness is lost as the eccentricity approaches touchdown. In contrast, high-speed plain bearings show improved stiffness at high eccentricity. At $K=9$, load support increases even further. However, caution should be exercised if there is any doubt whether perfect alignment can be achieved. Any misalignment will cause journal touchdown at reduced eccentricity ratio.

High-speed bearings may be designed for even higher loadings by increasing power ratio above the recommended range. Bearing loads at high K values and high eccentricity ratio are primarily hydrodynamic. As a rough guide, load support at high eccentricity ratio increases almost in proportion with $\sqrt{K}$ so that, for $K=12$, load support is almost double the load support at $K=3$. Bearings operated in the laboratory with values of power ratio in excess of 100 become less efficient on a load/power basis. There is also a danger of thermal collapse. It was found that power ratio greater than 1000 caused eccentricity ratio to increase, indicating incipient thermal collapse. Obviously speed was reduced before damage could occur.

Warning: Temperature rise increases with very high values of power ratio. With the double-row configuration a hot spot develops in the central region of the bearing near the point of minimum film thickness. This is due to "hot oil carry-over" around the bearing and back into the hot zone. The recommended design range, $3 \leq K \leq 9$, reduces risk of thermal problems by ensuring a healthy balance between hydrostatic and hydrodynamic lubricant flow. This ensures sufficient through-flow rather than excessive recirculatory flow around a bearing.

Length/Diameter Ratio

Load data are given in Figure 10.8 for $L/D = 0.5$ and $L/D = 2.0$. Hydrodynamic load support is significantly lower for $L/D = 0.5$. Also, bearing area is four times larger with $L/D = 2$ so load support is more than four times larger.

Land-Width Ratio

Effects of land-width ratio are demonstrated in Figures 10.7 and 10.8. Supported loads are a maximum when $a/L = 0.1$. However, load support starts to increase again as land-width ratio is reduced to the single-row value, $a/L = 0.5$. Single-row bearings have a particular advantage over double-row bearings for very high power ratio K. The direct outflow from single-row bearings is more effective in purging hot oil from the bearing under more extreme conditions, thus avoiding a hot spot.

Concentric Pressure Ratio

Effects of pressure ratio on high-speed load support are illustrated in Figure 10.9. Bearing load support increases with pressure ratio within the recommended range between 0.4 and 0.7, and tends to level off at higher values. While it is possible to design hybrid bearings for even higher values of pressure ratio, this is not generally recommended since the zero-speed load support diminishes down to zero. Figures 10.7−10.9 indicate a steep rise in load support for eccentricity ratio ε larger than 0.75. However, it is not recommended to operate at very high values of eccentricity, since it is difficult to ensure complete freedom from misalignment

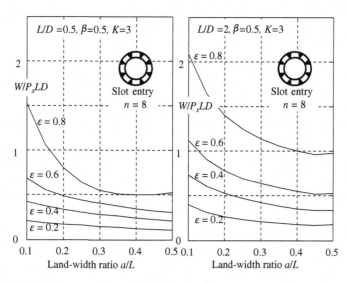

Figure 10.8: *W* vs. *a/L* for Eight-Slot Bearings: *L/D* = 0.5 and *L/D* = 2.

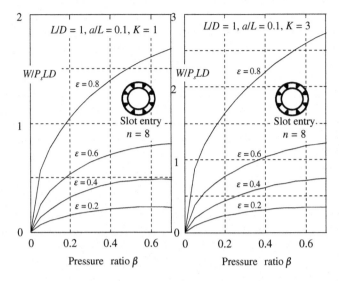

Figure 10.9: *W* vs. *β* for Eight-Slot Bearings: *K* = 1 and *K* = 3.

and allowance needs to be made for film thickness variations due to factors such as surface roughness and roundness deviations.

10.8 Hybrid Hole-Entry Bearings

It has already been shown (Figures 10.4 and 10.5) that 12-slot bearings offer better zero-speed load support than eight-slot bearings. Figure 10.10 gives load data for

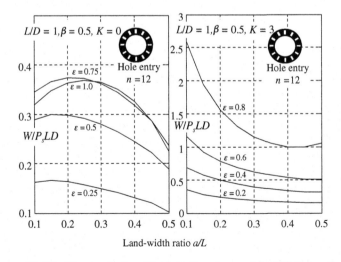

Figure 10.10: *W* vs. *a/L* for 12-Hole Bearings with 10° Entry Ports.

12-hole/row bearings at $K=0$ and at $K=3$. At zero speed, a 12-hole/row bearing gives better load support than an eight-hole/row bearing but is inferior to the eight-slot/row bearing.

At high speed, load support of a 12-hole/row bearing is comparable to an eight-slot/row bearing or a 12-slot bearing.

10.9 Size of Slot and Hole-Entry Ports

Load supported by slot and hole-entry bearings depends on the size of the entry ports. A small entry port gives less load support at zero speed than larger entry ports. This is a consequence of loss of pressure in the circumferential spacing between the holes. A convenient way to describe entry port size is by circumferential arc length of the ports, say 10°, 20°, 30° or 40°. For 12 slots per row and a slot-entry arrangement, the port length is 25°. For eight slots per row and a slot-entry arrangement, the port length is 40°. Hole-entry arrangements have smaller port lengths. Hole entry is used as shorthand to describe what may consist of circumferential grooves designed to spread port pressures more effectively than circular ports.

Zero-speed load support is reduced with 10° hole-entry ports compared with 25° entry ports for 12-slot/row bearings. The reduction is particularly apparent for $a/L=0.1$ (Figure 10.10). However, at high speed where $K=3$, load support is slightly better than for slot-entry bearings.

Increasing speeds to make $K=9$ or 12 (Figure 10.11) shows further increases in load support at high eccentricity, almost in proportion to $\sqrt{K}$.

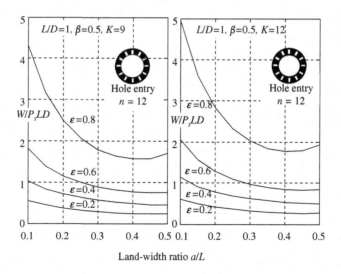

Figure 10.11: *W* vs. *a/L* for 12-Hole Bearings with 10° Entry Ports.

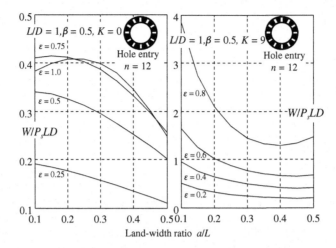

Figure 10.12: *W* vs. *a/L* for 12-Hole Bearings: 20° Entry Ports.

Zero-speed load support is increased with 20° entry ports (Figure 10.12) and is almost the same as for the eight-slot/row bearing. However, at higher speeds, $K = 9$ or 12, load support is slightly reduced for 20° ports compared with 10° ports.

Twenty-five-degree entry ports make a 12-hole/row bearing identical to a conventional 12-slot entry bearing.

10.10 Summary of Hydrostatic and Hybrid Load Support

- Hole entry is better than slot entry at high speeds but inferior at zero speed.
- Data for high-speed slot-entry bearings provide a conservative basis for design of high-speed hole-entry hybrid bearings and further charts are not therefore required to cover all possible design variations.
- Caution is necessary with regard to zero-speed load support with hole-entry bearings, since zero-speed load support is reduced.

10.11 Concentric Hydrostatic/Hybrid Flow

Flow data for the zero-speed concentric condition are presented in Figure 10.13 for slot- and hole-entry bearings with $L/D = 1$. Flow from a slot-entry bearing is slightly overestimated for the concentric condition from

$$\frac{q\eta}{P_s h_o^3} \approx \frac{\pi D}{6a} \cdot \beta \tag{10.4}$$

In selecting a pump, this figure should be increased by a factor, perhaps 40%, at the maximum pressure ratio for a slot-entry bearing to allow for increases in eccentricity ratio (see Figure 9.5b).

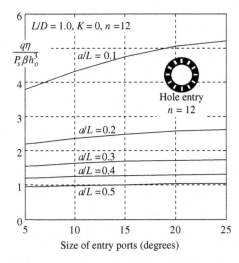

Figure 10.13: Concentric Flow: Slot and Hole-Entry Bearings, *n* = 12.

Flow from a hole-entry bearing is reduced at zero speed due to the restricted size of the entry ports. This is seen most strongly for thin-land bearings (Figures 10.13 and 10.14).

Flow from a slot-entry bearing can be read from the figures for a 25° entry port in the case of a 12-slot bearing or a 40° entry port for an eight-slot bearing.

Figure 10.15 gives flow for *L/D* ratios of 0.5 and 2.0. Although Figure 10.15 is for eight-slot and eight-hole bearings, flow for 12-slot and 12-hole bearings is not too different for proportionately sized entry ports.

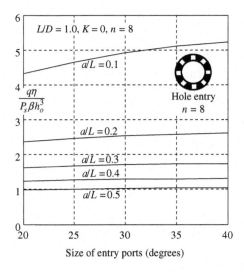

Figure 10.14: Concentric Flow: Slot- and Hole-Entry Bearings, *n* = 8.

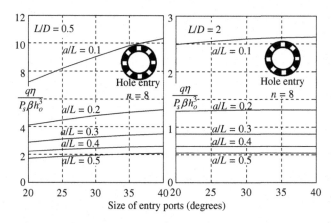

Figure 10.15: Concentric Flow: Slot and Hole Entry, $L/D = 0.5$ and 2.

10.12 Power and Temperature Rise

As in Chapter 3, expressions for pumping power and friction power are

$$H_p = P_s q \qquad \text{Pumping power} \tag{10.5}$$

$$H_f = \frac{\eta A_f U^2}{h} \qquad \text{power} \tag{10.6}$$

For plain bearings,

$$A_f = \pi D L \qquad \text{Friction area} \tag{10.7}$$

$$K = \frac{H_f}{H_p} \qquad \text{Power ratio} \tag{10.8}$$

Minimum Power

For *hydrostatic bearings* load support is defined at zero speed. Minimum power across the speed range is achieved by operating with $1 \leq K \leq 3$, as explained in Chapter 3.

For *hybrid bearings* maximum load is defined at speed. However, zero-speed load support of a hybrid bearing must be sufficient to support the maximum zero-speed load. A more economical bearing can be achieved by increasing the power ratio K. Load/power is substantially increased by operating with $3 \leq K \leq 9$.

Temperature Rise

Maximum temperature rise of the lubricant as it passes through the bearing is given by

$$\Delta T = \frac{P_s(1 + K)}{JC\rho} \tag{10.9}$$

Speed Parameter S_h (Use Optional)

The speed parameter is defined as

$$S_h = \frac{\eta N}{P_s}\left(\frac{D}{2h_o}\right)^2 \tag{10.10}$$

The parameter may be used, if desired, to select suitable combinations of viscosity, rotational speed in rev/s, supply pressure, bearing diameter, and bearing film thickness h_o. Values of S_h can be obtained from charts (Figure 10.3) or determined for other values of K from

$$S_h = \frac{1}{4\pi}\frac{D}{L}\sqrt{\frac{K\beta}{6a/L}}$$

For example, for $K = 3$, $L/D = 1$, $\beta = 0.4$, and $a/L = 0.1$:

$$S_h = \frac{1}{4\pi}\cdot 1 \cdot \sqrt{\frac{3 \times 0.4}{6 \times 0.1}} = 0.1125$$

so that $1/S_h = 8.89$, in agreement with Figure 10.3. It is interesting to note that $S_h = \overline{W}\cdot S$, where Sommerfeld number $S = (\eta N/Pav)(D/2h_o)^2$ is used in hydrodynamic bearing design and $\overline{W} = W/P_sLD$.

■ Example 10.1 Hybrid Hydrostatic Bearing

Design a suitable slot-entry hybrid bearing to carry the same load as in Example 9.4. The zero-speed load is less than 0.4 kN and maximum load is less than 4 kN (900 lbf) at 20 rev/s. Supply pressure P_s is reduced to 2 MN/m^2 (725 lbf/in^2) for the hybrid bearing.

Solution
From Figures 10.7 and 10.9 for $L/D = 1$, $\varepsilon = 0.6$, $a/L = 0.1$, $K = 3$:

$$\frac{W}{P_sLD} = 1.05 = \frac{4000}{2 \times 10^6 \times D^2}$$

so that $D = 43.6$ mm, say $D = 44$ mm (1.72 in), and $a = 0.1 \times 44 = 4.4$ mm (0.173 in).

Zero-speed load support may be checked within the range $\beta = 0.4-0.7$ and $\varepsilon = 0.5$ from Figure 10.6 for $K = 0$:

$$\frac{W}{P_sLD} = 0.27$$

so that $W = 0.27 \times 2 \times 10^6 \times 44^2 \times 10^{-6} = 1045$ N (235 lbf). This load support is more than double the 400 N required at zero speed.

From Figure 10.3, IT5 tolerance grade:

$$L \text{ limit} = 10.8 \times 10^{-4} : \quad 2h_o(L) = L \text{ limit} \times D = 10.8 \times 10^{-4} \times 44 \times 10^{-3}$$
$$= 47.5 \text{ μm } (0.00187 \text{ in})$$
$$U \text{ limit} = 1.5 \times L \text{ limit so that } 2h_o(U) = 1.5 \times 2h_o(L) = 71.25 \text{ μm } (0.0028 \text{ in})$$

$$\frac{P_s D}{\eta N L} = 3.12 \times 10^6 \text{ from Figure 10.3}$$

so that

$$\eta = \frac{2 \times 10^6}{3.12 \times 10^6 \times 20} = 0.032 \text{ N s/m}^2 \text{ or } 32 \text{ cP } (4.65 \times 10^{-6}\text{reyn})$$

Maximum flow corresponds to the maximum clearance condition when $K = 3$, $\beta = 0.4$, $h_o = 35.62$ μm. From equation (10.4) and values for D, P_s, and η:

$$q_o = \frac{P_s \beta h_o^3}{\eta} \frac{\pi D}{6a}$$

$$q_o = \frac{2 \times 10^6 \times 0.4 \times 35.62^3 \times 10^{-18}}{0.032} \times \frac{\pi \times 44}{6 \times 4.4} = 5.92 \times 10^{-6} \text{ m}^3/\text{s or } 0.355 \text{ l/min}$$

Minimum flow corresponds to the minimum clearance condition when $K = 9$, $\beta = 0.7$, $h_o = 23.75$ μm:

$$q_o = \frac{2 \times 10^6 \times 0.7 \times 23.75^3 \times 10^{-18}}{0.032} \times \frac{\pi \times 44}{6 \times 4.4} = 3.07 \times 10^{-6} \text{ m}^3/\text{s or } 0.184 \text{ l/min}$$

At $K = 3$: Pumping power $= P_s q = 2 \times 10^6 \times 5.92 \times 10^{-6} = 11.84 \text{ W}$

At $K = 9$: Pumping power $= P_s q = 2 \times 10^6 \times 3.07 \times 10^{-6} = 6.14 \text{ W}$

$$\text{Maximum friction power} = \frac{\eta A_f U^2}{h}$$

$$= \frac{0.032 \times \pi \times 44^2 \times 10^{-6} \times (\pi \times 44 \times 10^{-3} \times 20)^2}{23.75 \times 10^{-6}}$$

$$= 62.6 \text{ W}$$

Maximum total power $= 68.8 \text{ W}$ at $K = 9$

This may be compared with a total power of 127 W in Example 9.4. The plain bearing supporting the applied loads in hybrid mode represents a significant power saving.

Maximum temperature rise per pass $= 0.6 \times 10^{-6}(1 + K)P_s = 0.6 \times 10^{-6}(1 + 9) \times 2 \times 10^6 = 12 \,°\text{C at } K = 9$ compared with $4.8 \,°\text{C at } K = 3$. ∎

Part B Aerostatic Plain Journal Bearings

10.13 Introduction to Aerostatic Journal Bearings

Aerostatic journal bearings may be either hole entry or slot entry, as illustrated in Figures 10.1 and 10.2. With orifice restrictors, it is better for the gas to enter into the bearing from shallow and narrow slot-shaped pockets or grooves rather than from circular recesses. This is because entry pressure may be spread circumferentially while maintaining a small recess volume. Recess volumes should be small for good dynamic stability.

Effects of orifice restrictors on flow are more complex than for slot restrictors. Flow and load are both reduced with a small number of simple orifices or with short bearings, as described by Powell (1971). Restrictor effects on flow for aerostatic bearings are discussed in Chapter 5. Effects of recess volume and recess size are also discussed in Chapters 4 (Section 4.2) and 5.

Aerostatic bearings at very high speed have a hybrid action because aerodynamic load support is a direct consequence of speed. To achieve significant aerodynamic load support requires similar values of power ratio as for hydrostatic hybrid bearings. In practice, Powell (1971) showed that aerodynamic load support can increase load support by two or three times compared with aerostatic load support employing a single central row of six orifices. This increase was achieved with a small clearance and $L/D = 2$.

Radial bearing clearances may be slightly smaller for aerostatic bearings than for hydrostatic bearings. Smaller clearances reduce flow. Powell (1971) quotes a typical radial clearance of 25 μm for a 50-mm-diameter bearing, the same as indicated in Figure 10.3.

In general, similar principles apply for design of aerostatic bearings as for hydrostatic bearings.

10.14 Concentric Aerostatic Pressure Ratio, K_{go}

The concentric pressure ratio for most aerostatic bearings is recommended to lie within the range:

$$0.4 \leq K_{go} \leq 0.7 \tag{10.11}$$

This range maintains zero-speed load support near the maximum but also means better stability is ensured than with low values of concentric pressure ratio. Generally, lower values of K_{go} can be tolerated with slot restrictors than with orifice restrictors. Stability is discussed further in Chapter 5. For orifice-fed aerostatic bearings, it may be worthwhile reducing the range so that $0.5 < K_{go} < 0.7$. This reduces the clearance tolerance (see below) and reduces the risk of pneumatic instability.

Maximum clearance $h_o(U)$ corresponds to $K_{go} = 0.4$. Minimum clearance $h_o(L)$ corresponds to $K_{go} = 0.7$. Maximum load support is achieved when restrictors are adjusted so that $K_{go} \approx 0.5$.

10.15 Aerostatic Clearance and Clearance Limits

A bearing film thickness tolerance for most aerostatic journal bearings is discussed in Chapter 6. Application of a clearance tolerance is optional. It is recommended that $h_o(U) = 1.5h_o(L)$. For orifice feed, a reduced range $h_o(U) = 1.25h_o(L)$ corresponds to the reduced pressure ratio range $0.5 < K_{go} < 0.7$.

An alternative to preselecting clearance limits is to adjust restrictors after the journal and bearing have been manufactured and measured for size. Restrictors may then be manufactured subsequent to measurement of bearing clearance and adjusted to achieve the optimum pressure ratio $K_{go} = 0.5$.

10.16 Aerostatic Load Support

An indication of load support for aerostatic journal bearings is shown in Figure 10.16. Slot-entry bearings support higher load for short bearings. Also, bearing stability is more easily ensured for slot restrictors than for recessed orifice restrictors.

It can be seen that load support for aerostatic journal bearings follows similar trends to hydrostatic journal bearings. The main difference between orifice-fed bearings and slot-fed bearings is a reduction in load support with orifice feed due to loss of pressure in the region surrounding entry ports, particularly for $L/D \leq 1$. This is similar to findings for hydrostatic bearings.

Orifice-fed bearings offer higher stiffness at low eccentricity ratio, although concentric stiffness is often secondary compared to average stiffness. Any advantage of higher stiffness at low eccentricity is lost at high eccentricity. At high eccentricity there is also increased danger of "lock-up" with orifices due to loss of pressure around the entry ports in the minimum film thickness region. An experimental comparison between orifice-fed and slot-fed double-entry journal bearings is given by Stout et al. (1978).

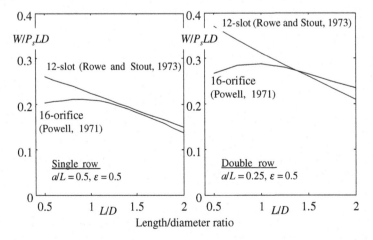

Figure 10.16: Aerostatic Load for Single-Row and Double-Row Aerostatic Bearings: $n = 12$ Slots and $n = 16$ Orifices per Row, $K_{go} = 0.5$, $p_s/p_a = 5$.

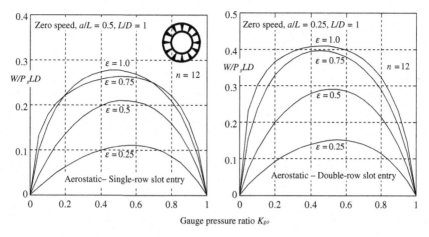

Figure 10.17: Aerostatic Loads: $p_s/p_a = 5$.

Figure 10.17 shows variation of load support with pressure ratio K_{go} for slot-entry bearings. Load values are slightly reduced compared with hydrostatic bearings for double-row bearings but are otherwise reasonably similar. For single-row bearings, values are marginally increased compared with hydrostatic bearings. The difference for double-row bearings is explained by effects of circumferential flow for aerostatic bearings.

A maximum applied load is approximately $0.25P_sLD$ for double-entry aerostatic bearings and $0.5 < L/D < 1$. Also, slot-entry bearings are preferred to orifice-fed bearings for $L/D = 0.5$.

For single-entry bearings, maximum applied load is approximately $0.2P_sLD$. These figures should be reduced for longer bearings where $1 < L/D < 2$, as shown by Figure 10.16.

10.17 Concentric Aerostatic Flow

Free air flow from a slot-entry bearing is slightly overestimated for the concentric condition from

$$\frac{q_o\eta}{P_sh_o^3} \approx (\frac{p_{ro} + p_a}{2p_a})\frac{\pi D}{6a} \cdot K_{go} \qquad (10.12)$$

Gauge inlet pressure is $P_{ro} = K_{go}P_s$ and absolute inlet pressure is $p_{ro} = P_{ro} + p_a$. Flow from a hole-entry bearing is reduced depending on the proportion of the circumference surrounded by the groove length or the pocket diameter or the hole-entry diameter. For example, if entry grooves surround 60% of the circumference, flow will lie in the range 60–100% of the value given by equation (10.12). Eighty percent is the mean of these two values and is a reasonable estimate for $L/D = 1$ and $a/L = 0.25$.

The flow required for restrictor design is the flow per restrictor. For a two-row bearing and 12 slots per row, the restrictor flow is

$$q_{restrictor} = \frac{q_{total}}{24}$$

■ *Example 10.2 Aerostatic Journal Bearing*

Design a suitable double-row aerostatic slot-entry bearing to carry a load of 0.4 kN (90 lbf) at 2400 rev/min (40 rev/s).

Given data:

Gauge supply pressure, P_s	0.414 MN/m^2 (60 lbf/in^2)
Atmospheric pressure, p_a	0.101$_5$ MN/m^2 (14.7 lbf/in^2)
Viscosity, η	18.2 × 10^{-6} N s/m^2 (2.64 × 10^{-9} reyn)
Gauge pressure ratio, K_{go}	0.5

Solution

An approximate guide from Section 10.16 for $L/D = 1$, $\varepsilon = 0.5$, $a/L = 0.25$, and $K_{go} = 0.5$ is

$$\frac{W}{P_s L D} = 0.25 = \frac{400}{0.41 \times 10^6 \times D^2}$$

so that $D = 62.5$ mm (2.46 in).

From Figure 10.17, the suitability of this design may be checked across the range $K_{go} = 0.4–0.7$ and $\varepsilon = 0.5$:

$$\frac{W}{P_s L D} = 0.275 \text{ at } K_{go} = 0.4 \text{ reducing to } 0.26 \text{ at } K_{go} = 0.7$$

Land-width ratio $a/L = 0.25$ so that land width $a = 0.25 \times 62.5 = 15.62$ mm (0.4 in)

From Figure 10.3, IT5 tolerance grade leads to

$$2h_o(L) = L \text{ limit} \times D = 9 \times 10^{-4} \times 62.5 \times 10^{-3} = 56.2 \text{ μm (0.0022 in)}$$
$$2h_o(U) = 1.5 \times 2h_o(L) = 84.3 \text{ μm (0.0033 in)}$$

Maximum flow corresponds to maximum clearance when $K_{go} = 0.4$ and $h_o = 42.2$ μm. Flow is given by equation (10.12):

$$\frac{q_o \eta}{P_s h_o^3} \approx \left(\frac{p_{ro} + p_a}{2p_a}\right) \frac{\pi D}{6a} \cdot K_{go}$$

$$p_{ro} + p_a = K_{go} \cdot P_s + 2p_a = 0.207 + 0.202 = 0.409 \text{ MN/m}^2$$

$$q_o = \frac{0.414 \times 10^6 \times 42.2^3 \times 10^{-18}}{18.2 \times 10^{-6}} \times \left(\frac{0.409 \times 10^6}{2 \times 0.101 \times 10^6}\right) \times \frac{\pi 62.5 \times 10^{-3}}{6 \times 15.62 \times 10^{-3}} \times 0.4$$

$$= 0.0029 \text{ m}^3/\text{s (174 l/min) free air flow}$$

If this air flow is considered rather high, flow can be halved by reducing the maximum clearance by 20% for the same pressure ratio.

Minimum flow corresponds to minimum clearance when $K_{go} = 0.7$ and $h_o = 28.1$ µm:

$$q_o = \frac{0.414 \times 10^6 \times 28.1^3 \times 10^{-18}}{18.2 \times 10^{-6}} \times \left(\frac{0.409 \times 10^6}{2 \times 0.101 \times 10^6}\right) \times \frac{\pi 62.5 \times 10^{-3}}{6 \times 15.62 \times 10^{-3}} \times 0.7$$

$$= 0.0015 \text{ m}^3/\text{s} \ (90 \text{ l/min})$$

From equation (3.1b), pumping power at maximum clearance is

$$H_p = p_a q_a \log_e\left(\frac{p_s}{p_a}\right) = 0.101 \times 10^6 \times 0.0029 \times \log_e\left(\frac{0.515}{0.101}\right) = 471 \text{ W}$$

Pumping power at minimum clearance is

$$H_p = p_a q_a \log_e\left(\frac{p_s}{p_a}\right) = 0.101 \times 10^6 \times 0.0015 \times \log_e\left(\frac{0.515}{0.101}\right) = 244 \text{ W}$$

Friction power at minimum clearance is

$$\frac{\eta A_f U^2}{h} = \frac{18.2 \times 10^{-6} \times \pi \times 62.5^2 \times 10^{-6} \times (\pi \times 62.5 \times 10^{-3} \times 40)^2}{28.1 \times 10^{-6}} = 0.49 \text{ W}$$

Power ratio $K = 0.49/244 = 0.002$ at minimum clearance.

The total power of approximately 244 W may be compared with 69 W in Example 10.1 and 124 W in Example 9.4. The plain bearing supports larger loads in hybrid hydrostatic mode but the aerostatic bearing runs at twice the speed with negligible friction power.

In this example the power ratio is extremely low and therefore hybrid support is negligible. Further increasing speed to 5000 rev/min in this example and reducing clearance by 75% would bring the bearing into the hybrid region. A smaller bearing could be employed and power loss reduced.

■

10.18 Hybrid Aerostatic Journal Bearings

Hybrid aerostatic bearings achieve increased load support at high bearing speeds with very small clearances. Load support for a hybrid bearing depends on the hybrid compressibility number Λ_h (pronounced lambda). The hybrid compressibility number is defined as

$$\Lambda_h = \frac{\eta \omega}{p_m}\left(\frac{D}{2h_o}\right)^2$$

where the mean pressure $p_m = (p_s + p_a)/2$. Also, η is the viscosity of the gas, ω is the angular shaft speed defined in radians per second, D is the shaft diameter, and $2h_o$ is the diametral clearance. For hybrid aerostatic journal bearings $D/2h_o$ is typically of the order of 2.5×10^3.

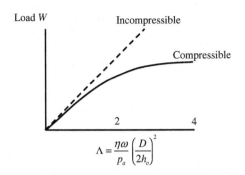

Figure 10.18: Typical Variation of Load Support with Compressibility Number Λ in an Aerodynamic Journal Bearing.

Powell (1971) points out for a purely aerodynamic bearing that load support increases linearly with compressibility number up to values of approximately 0.15, but levels off and becomes virtually independent of speed above $\Lambda = 4$. This is shown schematically in Figure 10.18.

Powell (1971) demonstrated for typical orifice-fed journal bearings with central admission and $L/D = 0.8$ that load support can be approximately doubled at a compressibility number of $\Lambda_h = 0.33$. For $L/D = 2$, Powell demonstrated an approximate doubling of load support with $\Lambda_h = 0.13$.

Figure 10.19 gives hybrid load data for $K = 3$ and $L/D = 1$ and $L/D = 2$. Comparison with previous charts for hydrostatic bearings shows that hybrid aerostatic load support is reduced for $a/L = 0.1$ by less than 30% but improved for bearings with central admission. Overall, the trends are remarkably similar. It will be observed from Figure 10.16 that load support is

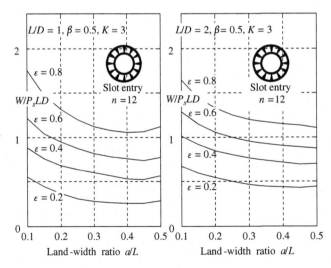

Figure 10.19: W vs. a/L for 12 Slots/Row Hybrid Aerostatic Bearings with $p_s/p_a = 5$: $L/D = 1$ and $L/D = 2$.

reasonably constant for L/D in the region $L/D = 1$. This allows the range of application of the data to be extended using a degree of caution.

The above data indicate speeds and clearances required for hybrid load support. These findings are verified using the power ratio approach employed throughout this book and reveal the unique usefulness of K for simpler practical design.

10.19 Materials and Surface Texture for Journal Bearings

Bearings for high loadings and eccentricity ratios should be designed with particularly careful consideration of bearing materials and surface texture. At high eccentricity ratio, minimum film thickness should always be greater than the surface roughness by a factor much greater than 2. The material chosen for the bearing should be strong enough to withstand the maximum pressures and it is wise to investigate these pressures very carefully if a soft bearing liner is employed. For accuracy and rigidity, a suitably selected wear-resistant phosphor bronze material for the bearing may be employed together with a hardened and ground nitriding-steel journal. This combination should be adequate for most hydrostatic applications.

For aerostatic applications, even more care needs to be taken to ensure corrosion-resistant and wear-resistant surfaces. The journal may be given a hard-chrome surface for improved corrosion resistance. The bearing may typically be a suitable phosphor bronze or stainless steel. It is recommended that specialist advice be sought.

References

Ives, D., & Rowe, W. B. (1987). The effect of multiple supply sources on the performance of heavily loaded pressurized high-speed journal bearings. *Proceedings of the Institution of Mechanical Engineers, 121*−127, Paper C199/87.

Powell, J. W. (1971). *Design of Aerostatic Bearings*. Machinery Publishing Company Ltd.

Rowe, W. B., & Koshal, D. (1980). A new basis for the optimisation of hybrid journal bearings. *Wear, 64*(3, Oct), 115−131.

Rowe, W. B., & Stout, K. J. (1973). Design of externally pressurized gas fed bearings employing slot restrictors. *Tribology International, 140*−144, August.

Rowe, W. B., Koshal, D., & Stout, K. J. (1976). Slot entry bearings for hybrid hydrodynamic and hydrostatic operation. *Journal of Mechanical Engineering Science, Proceedings of the Institution of Mechanical Engineers, 18*(2), 73−78.

Rowe, W. B., Koshal, D., & Stout, K. J. (1977). Investigations on recessed hydrostatic and slot-entry journal bearings for hybrid hydrodynamic and hydrostatic operation. *Wear, 43*(1), 55−70.

Rowe, W. B., Chu, S. X., Chong, F. S., & Weston, W. (1982). Hybrid journal bearings − with particular reference to hole-entry configurations. *Tribology International, 15*(6), 339−348.

Shires, G. L., & Dee, C. W. (1967). Pressurized bearings with inlet slots. *Proceedings of the Southampton Gas Bearing Symposium*. Paper 7.

Stout, K. J., Pink, E. G., & Tawfik, M. (1978). Comparison of slot-entry and orifice-compensated gas journal bearings. *Wear, 51*, 137−145.

Yoshimoto, S., Rowe, W. B., & Ives, D. (1988). A theoretical investigation of the effect of inlet pocket size on the performance of hole-entry hybrid journal bearings employing capillary restrictors. *Wear, 127*, 307−318.

Appendix: Tabular Design Procedures

Procedure 10.A1 Hybrid Hydrostatic Journal Bearings

Step	Symbol	Description of Operation	Notes
1	W	Specify: zero-speed load	From design requirements
2	P_s	Specify: supply pressure	Suggestion: 2 MN/m^2 (291 lbf/in^2)
3	n	Specify: number of grooves/slots	Suggestion: 2 rows of 8–12 slots
4	β	Specify: pressure ratio, P_{ro}/P_s	Suggestion: 0.5
5	L/D	Specify: length/diameter	Suggestion: 1
6	a/L	Specify: land-width ratio	Suggestion: 0.25 for hydrostatic or 0.1 for hybrid
7	θ	Specify: bridge angle	Suggestion: 5° for $n = 12$
8	$\overline{W}$	Read: zero-speed load factor	See charts
9	$\overline{B}$	Calculate: flow factor	From $(\pi D/L)/(6a/L)$
10	D	Calculate: from $\sqrt{\dfrac{W}{P_s \overline{W} L/D}}$	If diameter D is too large, increase P_s
11	L	Calculate: length	From $D \times L/D$
12	a	Calculate: land width	From land-width ratio
13	A_f	Calculate: friction area	From πDL
14	K	Specify: max power ratio	Suggest $K = 3$ for hydrostatic or 9 for hybrid
15	h_o	Specify: min radial bearing gap	h_o should be at least 5–10 times flatness tolerance
16	U	Calculate: sliding speed	From πDN
17	η	Calculate: viscosity	From $\eta = \dfrac{P_s h_o^2}{U}\sqrt{\dfrac{K\beta\overline{B}}{A_f}}$
18	q_0	Calculate: flow	$q_0 = \dfrac{P_s h_o^3 \beta \overline{B}}{\eta}$
19	$\overline{W}$	Read/estimate: hybrid load factor	See charts for approximate value
20	W	Calculate: hybrid load	From $P_s L D \overline{W}$
21	H_p	Calculate: pumping power	From $H_p = P_s q$
22	H_f	Calculate: friction power	From $H_f = \eta A_f U^2/h_o$
23	ΔT	Calculate: temperature rise/pass	From $0.6 \times 10^{-6}(1 + K)P_s$

Example 10.A1 Hybrid Journal Bearing for 5000 N at Zero Speed and 10,000 N at 2000 rev/min

Step	Symbol	Example of Working	Result
1	W	Zero-speed load	5000 N (1124 lbf)
2	P_s	Supply pressure	$P_s = 2$ MN/m^2
3	n	Two rows of 12 slots	12
4	β	Pressure ratio	0.5
5	L/D	Length/diameter	1.0
6	a/L	Land-width ratio	0.1
7	θ	Bridge angle	5°
8	$\overline{W}$	Load factor from Figure 10.6	0.32
9	$\overline{B}$	$(\pi \times 1)/(60 \times 0.25)$	2.094
10	D	$\sqrt{\dfrac{5000}{2 \times 10^6 \times 0.32 \times 1}}$ $\dfrac{}{0.0884 \times 1}$	0.0884 m(3.480in)
11	L		0.0884 m (3.48 in)
12	a	0.1218×0.25	0.03045 m (1.2 in)
13	A_f	$\pi \times 0.0884^2$	0.02455 m^2 (38 in^2)
14	K	Power ratio	9
15	h_o	Minimum radial bearing gap	31 µm (0.00122 in)
16	U	$\pi \times 0.0884 \times 2000/60$	9.257 m/s
17	η	$\dfrac{2 \times 10^6 \times 31^2 \times 10^{-12}}{9.257} \sqrt{\dfrac{9 \times 0.5 \times 2.094}{0.02455}}$	0.004068 N s/m^2 (4.068 cP)
18	q_o	$\dfrac{2 \times 10^6 \times 31^3 \times 10^{-18} \times 0.5 \times 2.094}{0.004068}$	15.33×10^{-6} m^3/s (0.92 l/min)
19	$\overline{W}$	High-speed load factor from Figure 10.7	≈ 1.5 at $\varepsilon = 0.6$
20	W	$2 \times 10^6 \times 0.0884 \times 0.0884 \times 1.5$	23.44 kN (5270 lbf)
21	H_p	$2 \times 10^6 \times 15.33 \times 10^{-6}$	30.66 W
22	H_f	$0.004068 \times 0.02455 \times 9.257^2/31 \times 10^{-6}$	276 W
23	ΔT	$0.6 \times 10^6 (1+9) \times 2 \times 10^6$	12 °C

Procedure 10.A2 Aerostatic Journal Bearings

Step	Symbol	Description of Operation	Notes
1	W	Specify: extreme load	From design requirements
2	P_s, p_s	Specify: supply pressure (gauge and absolute)	Suggestion: $P_s = 0.45$ MN/m^2, $p_s = 0.551$ MN/m^2
3	n	Specify: number of holes/slots	Suggestion: two rows of 12 slots
4	K_{go}	Specify: pressure ratio, P_{ro}/P_s	Suggestion: 0.5
5	p_{ro}	Calculate: entry pressure	From $K_{go}P_s + p_a$
6	L/D	Specify: length/diameter	Suggestion: 1
7	a/L	Specify: land-width ratio	Suggestion: 0.25 for double entry
8	$\overline{W}$	Read or estimate: load factor	See Figure 10.17
9	$\overline{B}$	Calculate: flow shape factor	From $(\pi D/L)/(6a/L)$
10	D	Calculate: from $\sqrt{\dfrac{W}{P_s\overline{W}L/D}}$	If diameter D is too large, increase P_s
11	L	Calculate: length	From $D \times L/D$
12	a	Calculate: land width	From land-width ratio
13	h_o	Specify: radial bearing gap	h_o should be at least 5–10 times flatness tolerance
14	η	Specify: viscosity of gas	
15	q_o	Calculate: flow	$q_o = \dfrac{P_s h_o^3 K_{go}\overline{B}}{\eta}\dfrac{P_{ro}+P_a}{2P_a}$
16	λ	Calculate: average stiffness	From $\lambda \approx 2W/h_o$
17	A_f	Calculate: friction area	From πDL
18	H_p	Calculate: pumping power	From $H_p = p_a q_a \ln(p_s/p_a)$
19	H_f	Calculate: friction power	From $H_f = \dfrac{\eta A_f U^2}{h_o}$
20	K	Power ratio	From H_f/H_p

Example 10.A2 Aerostatic Plain Journal Bearings, Slot Entry, Sliding Speed 15 m/s, Maximum Applied Load 1000 N

Step	Symbol	Example of Working	Result
1	W	Extreme load	1000 N (225 lbf)
2	P_s, p_s	Gauge and absolute supply pressure	$P_s = 0.45$ MN/m^2
			$p_s = 0.551$ MN/m^2
3	n	Double entry and 12 slots/row	12
4	K_{go}	Pressure ratio	0.5
5	p_{ro}	$(0.5 \times 0.45) + 0.101 \times 10^6$	0.326 MN/m^2
6	L/D	Length/diameter	1
7	a/L	Land-width ratio	0.25
8	$\overline{W}$	Load factor	0.28
9	$\overline{B}$	$(\pi \times 1)/(6 \times 0.25)$	2.094
10	D	$\sqrt{\dfrac{1000}{0.45 \times 10^6 \times 0.28 \times 1}}$	0.0891 m (3.51 in)
11	L	0.0891×1	0.0891 m (3.51 in)
12	a	0.0891×0.25	0.0223 m (0.877 in)
13	h_o	Radial bearing gap	25 µm (0.001 in)
14	η	Viscosity of air	18.3×10^{-6} N s/m^2
15	q_o	$\dfrac{0.45 \times 10^6 \times 25^3 \times 10^{-18} \times 0.5 \times 2.094}{0.0000183} \dfrac{0.326 + 0.101}{2 \times 0.101}$	850.4×10^{-6} m^3/s (51 l/min)
16	λ	$\dfrac{2 \times 1000}{25 \times 10^{-6}}$	80 MN/m (0.457×10^6 lbf/in)
17	A_f	$\pi \times 0.0891 \times 0.0891$	0.02494 m^2 (38.66 in^2)
18	H_p	$0.101 \times 10^6 \times 850.4 \times 10^{-6} \times \ln(0.551/0.101)$	145.7 W
19	H_f	$\dfrac{18.3 \times 10^{-6} \times 0.0294 \times 15^2}{25 \times 10^{-6}}$	4.842 W
20	K	$4.842/145.7$	0.033

Procedure 10.A3 Hybrid Aerostatic Journal Bearings

Step	Symbol	Description of Operation	Notes
1	W	Specify: zero-speed load	From design requirements
2	P_s, p_s	Specify: supply pressure (gauge and absolute)	Suggestion: $P_s = 0.45$ MN/m^2 (65lbf/in^2) $p_s = 0.551$ MN/m^2
3	n	Specify: number of holes/slots	Suggestion: two rows of 12 slots
4	K_{go}	Specify: pressure ratio, P_{ro}/P_s	Suggestion: 0.5
5	p_{ro}	Calculate: entry pressure	From $K_{go}P_s + p_a$
6	L/D	Specify: length/diameter	Suggestion: 1
7	a/L	Specify: land-width ratio	Suggestion: 0.25 for double entry
8	$\overline{W}$	Read or estimate: load factor	See Figure 10.18
9	$\overline{B}$	Calculate: flow shape factor	From $(\pi D/L)/(6a/L)$
10	D	Calculate: from $\sqrt{\dfrac{W}{P_s\overline{W}L/D}}$	If diameter D is too large, increase P_s
11	L	Calculate: length	From $D \times L/D$
12	a	Calculate: land width	From land-width ratio
13	A_f	Calculate: friction area	From πDL
14	K	Specify: power ratio	Suggest $K = 3$
15	η	Specify: viscosity of gas	
16	h_o	Calculate:	$\left[\dfrac{\eta U}{P_s}\sqrt{\dfrac{2A_f}{K \cdot K_{go} \cdot \overline{B}}\dfrac{P_s}{(p_{ro}+p_a)} \cdot \dfrac{1}{\ln(p_s/p_a)}}\right]^{1/2}$
17	q_o	Calculate: flow	$q_o = \dfrac{P_s h_o^3 K_{go}\overline{B}}{\eta}\dfrac{p_{ro}+p_a}{2p_a}$
18	$\overline{W}$	Read/estimate: hybrid load factor	See charts for approximate value
19	W	Calculate: high-speed load	From $P_s LD\overline{W}$
20	H_p	Calculate: pumping power	From $H_p = p_a q_a \ln(p_s/p_a)$
21	H_f	Calculate: friction power	From $H_f = \dfrac{\eta A_f U^2}{h_o}$

Example 10.A3 Hybrid Aerostatic Plain Journal Bearings, Slot Entry, Sliding Speed 150 m/s, Maximum Applied Load 1.0 N at Zero Speed and 2.5 N at Speed

Step	Symbol	Example of Working	Result
1	W	Zero-speed load	1.0 N (0.404 lbf)
2	P_s, p_s	Gauge and absolute supply pressure	$P_s = 0.345$ MN/m^2 (50 lbf/in^2) $p_s = 0.446$ MN/m^2 (64.7 lbf/in^2)
3	n	Single entry and 8 slots/row	12
4	K_{go}	Pressure ratio	0.5
5	p_{ro}	$(0.5 \times 0.345 + 0.101) \times 10^6$	0.2735 MN/m^2
6	L/D	Length/diameter	0.8
7	a/L	Land-width ratio	0.5
8	$\overline{W}$	Load factor	0.2
9	$\overline{B}$	$(\pi \times 1/0.8)/(6 \times 0.5)$	1.309
10	D	$\sqrt{\dfrac{1.0}{0.345 \times 10^6 \times 0.2 \times 0.8}}$	0.00425 m, say 0.0047 m (0.185 in)
11	L	0.0047×0.8	0.0036 m (0.148 in)
12	a	0.0036×0.5	0.0018 m (0.074 in)
13	A_f	$\pi \times 0.0047 \times 0.0036$	53.16×10^{-6} m^2 (0.08239 in^2)
14	K	Power ratio	3
15	η	Viscosity of air	18.3×10^{-6} N s/m^2
16	h_o	$\left[\dfrac{18.3 \times 10^{-6} \times 150}{345,000} \sqrt{\dfrac{2 \times 53.22 \times 10^{-6}}{3 \times 0.5 \times 1.31} \dfrac{0.345}{(0.375)} \times \dfrac{1}{\ln(4.416)}}\right]^{1/2}$	6.79 μm
17	q_o	$\dfrac{0.345 \times 10^6 \times 6.79^3 \times 10^{-18} \times 0.5 \times 1.31}{0.0000183} \dfrac{0.2735 + 0.101}{2 \times 0.101}$	7.17×10^{-6} m^3/s (0.43 l/min)
18	$\overline{W}$	From Figure 10.19	≈ 0.6
19	W	$0.345 \times 10^6 \times 0.0036 \times 0.0047 \times 0.6$	3.5 N
20	H_p	$0.101 \times 10^6 \times 7.17 \times 10^{-6} \times \ln(0.446/0.101)$	1.08 W
21	H_f	$\dfrac{18.3 \times 10^{-6} \times 53.16 \times 10^{-6} \times 150^2}{6.79 \times 10^{-6}}$	3.22 W

CHAPTER 11

The Yates Bearing

Summary of Key Design Formulae

$$q_o = 2 \cdot \frac{P_s h_a^3}{\eta} \cdot \beta_2 \overline{B}_t \qquad \text{Hydrostatic flow}$$

$$q_o = 2 \cdot \frac{P_s h_a^3}{\eta} \cdot K_{go2} \overline{B}_t \cdot \left(\frac{p_2 + p_a}{2 p_a} \right) \qquad \text{Aerostatic flow}$$

$$W = P_s LD \cdot \overline{W} \qquad \text{Radial load support}$$

$$T = P_s A_{et} \cdot \overline{T} \qquad \text{Thrust pad load support}$$

$$A_{et} = A_{e1} + A_{e2} \qquad \text{Thrust pad effective area}$$

$$\overline{D} = \frac{D_o}{D_i} \qquad \text{Thrust pad diameter ratio}$$

$$\varepsilon_a = \frac{e_a}{h_a} \qquad \text{Axial eccentricity ratio}$$

$$\varepsilon_r = \frac{e_r}{h_j} \qquad \text{Radial eccentricity ratio}$$

$$H_f = \frac{\eta N^2 D^4}{2 h_j} \cdot \overline{F} \qquad \text{Friction power}$$

11.1 Introduction

Four arrangements are available to carry combined axial and radial loads. These are illustrated in Figures 1.8 and 1.9:

1. Separate thrust and journal bearings
2. Conical bearings
3. Spherical bearing
4. Combined journal and thrust bearings known as "Yates bearings".

Hydrostatic, Aerostatic and Hybrid Bearing Design.
DOI: 10.1016/B978-0-12-396994-1.00011-5

Although the Yates bearing illustrated has a recessed journal, the same principle applies if the journal bearing is a plain bearing of a type described in the previous chapter. This allows a Yates bearing to employ a plain hybrid journal bearing or a plain aerostatic bearing. The same data can be used.

A Yates-type bearing was first described by Yates (1950). The principle is to supply thrust faces by leakage flow from the journal bearing. Lund (1963) incorporated a pocket in the thrust face, as in Figure 11.1. The development was carried out by Mechanical Technology Inc. for the National Aeronautics and Space Administration, USA. A Yates bearing without recesses in the thrust faces is shown in Figure 11.2. Advantages of a Yates bearing were noted as being:

1. Pumping power per unit load is less than for a separately fed system (50%).
2. Friction in the thrust faces is less than for a separately fed thrust bearing.
3. Substantial thrust loads can be carried at a sacrifice of a proportion of radial load support. For gas bearings, axial load capacity is slightly greater than for a conventional thrust bearing. This statement could also be true in the case of liquid pressurized bearings but is achieved at the cost of a loss in axial dynamic stiffness and a reduction in journal bearing performance.
4. The thrust bearings have no feed system, thus the bearing arrangement is simplified, the cost of machining the end faces is reduced, and the space requirements are less. Moreover, simplification of the system should increase reliability.

A design procedure was provided for the Yates bearing by Wearing et al. (1970). Additional symbols and definitions are necessary for Yates bearings as indicated above and in

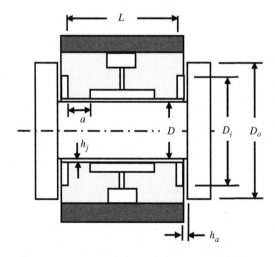

Figure 11.1: Geometry of the Yates Bearing.

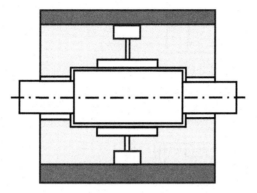

Figure 11.2: An Alternative Yates Bearing.

Figure 11.1 to distinguish journal bearing parameters from thrust bearing parameters. In all cases, $L/D = 1$.

For aerostatic versions, recesses should be minimized to a very small value, as described in previous chapters (see Chapter 5). Slot-entry bearings are likely to be more stable than pocketed orifice-fed bearings.

11.2 Principle of Operation

The Yates bearing principle is illustrated in Figure 11.3 by reference to pressure distributions and recess pressures.

Zero-load conditions are shown in Figure 11.3a. Journal recess pressures are set at approximately 60% of supply pressure. The thrust pad recess pressure is approximately 20% of supply pressure. Journal recesses are fed through flow restrictors. Capillary or slot-entry control has a special advantage for this application because the journal bearing forms part of the flow control to the thrust bearing and is by its nature a viscous restrictor. Fluid flows axially across the journal lands into the thrust pad recesses and then radially outwards across the thrust lands to exhaust at atmospheric pressure.

When an axial load T is applied, as in Figure 11.3b, the increased flow resistance on the loaded side causes the pressure to rise while the lessened resistance at the opposite face causes the pressure to reduce. The difference in thrust face pressures supports the axial load. When a radial load W is applied, pressures in the journal recesses vary but thrust pressures are almost constant, as shown in Figure 11.3c.

Typical variations in recess pressures are shown in Figure 11.4. Axial load variations are indicated in Figure 11.4a and radial load variations in Figure 11.4b.

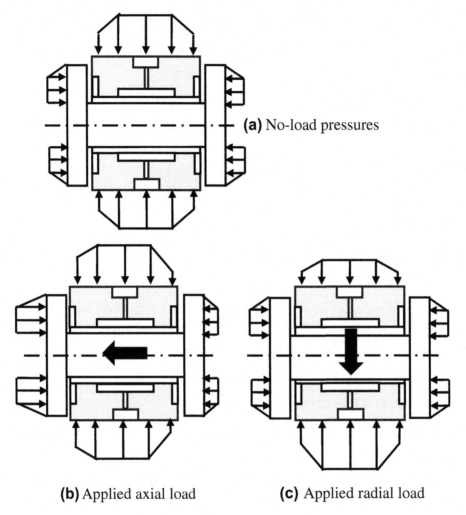

(a) No-load pressures

(b) Applied axial load   **(c)** Applied radial load

Figure 11.3: Effect of Axial and Radial Loads on Pressures.

11.3 Basic Parameters for the Yates Bearing

Concentric Gauge Recess Pressures

The concentric gauge pressure ratio at entry into the journal bearing is termed $\beta_1 = K_{go1} = P_1/P_s$ and in the thrust bearings is $\beta_2 = K_{go2} = P_2/P_s$, where P_2 is the gauge pressure at entry to the thrust pad and P_s is the supply gauge pressure. Splitting the pressure differences equally from supply pressure P_s down to $\beta_1 P_s$ and then down to $\beta_2 P_s$ allows bearing pressures to vary up and down equally in both the radial and axial directions. An equal split makes pressure ratios $\beta_1 = 0.67$ and $\beta_2 = 0.33$. A slightly larger maximum radial load support can be

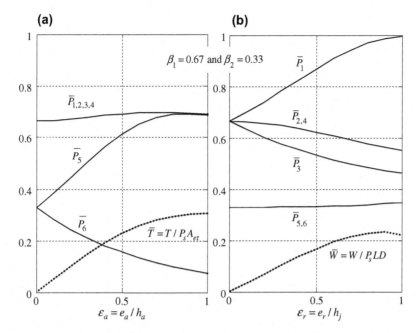

Figure 11.4: Effect of Axial and Radial Loads on Recess Pressures.
(a) Recess pressures and axial thrust with axial eccentricity. (b) Recess pressures and radial load with radial eccentricity.

obtained by reducing β_2 down to say 0.2. A general guide for β_1 may be obtained by analogy with a conventional hydrostatic bearing where the recommended pressure ratio is 0.5. For a Yates bearing it is recommended that

$$\beta_1 = (1 + \beta_2)/2 \qquad \text{Pressure ratios} \qquad (11.1)$$

The design procedure that follows assumes that this recommendation is followed. Thus, if the axial pressure ratio $\beta_2 = 0.2$, the journal pressure ratio is $\beta_1 = 0.6$. Adequate thrust bearing load support can usually be obtained with these values. The combination $\beta_1 = 0.6$ and $\beta_2 = 0.2$ will often, therefore, be adopted.

Flow

Total Flow

Flow q through a Yates bearing is the total flow and is therefore the sum of the flows through the two thrust pads. It is important to keep this in mind to avoid mistakes in balancing flow resistances. The design procedure below takes account of this requirement.

Thrust Bearing Flow

Concentric flow is the combined flow for both thrust pads. This gives rise to a factor 2 in Equations (11.2) and (11.3):

$$q = 2 \cdot \frac{P_s h_a^3}{\eta} \beta_2 \cdot \overline{B}_t \qquad \text{Hydrostatic} \qquad (11.2)$$

$$q = 2 \cdot \frac{P_s h_a^3}{\eta} K_{go2} \cdot \overline{B}_t \cdot \frac{p_2 + p_a}{2 p_a} \qquad \text{Aerostatic} \qquad (11.3)$$

The flow shape factor for the thrust pads $\overline{B}_t$ is given in Figure 11.5 or by

$$\overline{B}_t = \frac{\pi}{6 \ln(D_o / D_i)} \qquad (11.4)$$

Journal Bearing Flow

Concentric journal flow is

$$q = \frac{P_s h_j^3}{\eta} (\beta_1 - \beta_2) \cdot \overline{B}_j \qquad \text{Hydrostatic} \qquad (11.5)$$

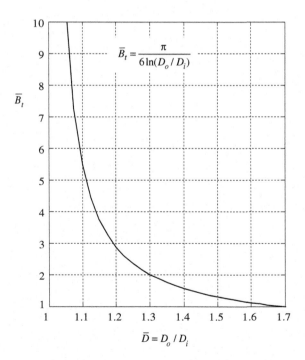

Figure 11.5: Flow Shape Factor for a Single Circular Thrust Pad.

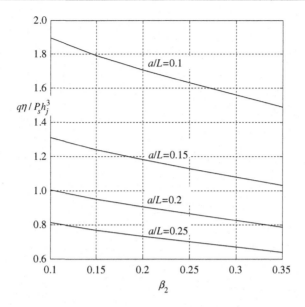

Figure 11.6: Hydrostatic Flow: for $n = 4$, $\theta = 30$, and $\beta_1 = (1 + \beta_2)/2$.

$$q = \frac{P_s h_j^3}{\eta}(K_{go1} - K_{go2}) \cdot \overline{B}_j \cdot \frac{p_1 + p_2}{2p_a} \qquad \text{Aerostatic} \qquad (11.6)$$

Flow shape factors $\overline{B}_j$ are given in Chapters 9 and 10 or the maximum concentric value by

$$\overline{B}_j = \frac{\pi D}{6a} \qquad (11.7)$$

The shape factors apply for both recessed and plain slot-entry bearings. With $a/L = 0.25$ and $L/D = 1$, $\overline{B}_j = 2.094$ so that the combination $\beta_1 = 0.6$ and $\beta_2 = 0.2$ for hydrostatic flow leads to $\overline{Q} = 0.4 \times 2.094 = 0.838$.

Hydrostatic flow for radial and axial eccentricity ratios of 0.5 is given in Figure 11.6 for several values of land-width ratio a/L. Aerostatic free air flow is increased compared with hydrostatic flows in Figure 11.6 by the factor $(p_1 + p_2)/2p_a$, where these are absolute pressures.

Flow reduces with increasing thrust bearing pressure ratio. The journal bearing pressure ratio is set according to the recommendation that $\beta_1 = (1 + \beta_2)/2$.

Restrictor Flow

Flow through a restrictor is

$$q_{\text{restrictor}} = \frac{q}{n} \qquad \text{for a single row of } n \text{ restrictors per row}$$

$$q_{\text{restrictor}} = \frac{q}{2n} \qquad \text{for a double row of } n \text{ restrictors per row}$$

Combined Radial and Thrust Loads

Combined load characteristics for a typical Yates bearing are shown in Figure 11.7. As can be seen from this graph, there is reduced radial load support with increasing thrust load and similarly thrust load support is reduced with increasing radial load.

Axial load support on the thrust bearings is primarily dependent on axial eccentricity ratio ε_a, where eccentricity ratios are defined in the usual way as in Figure 11.4. Maximum axial load support for a Yates bearing is roughly proportional to load support for conventional thrust bearings according to

$$T_{Yates} \approx \beta_1 \cdot T_{conventional}$$

Radial load support on the journal is primarily dependent on radial eccentricity ratio ε_r. Maximum radial load support for a Yates bearing is roughly proportional to load support for conventional journal bearings according to

$$W_{Yates} \approx (1 - \beta_2) \cdot W_{conventional}$$

These simple rules allow data for conventional recessed and plain hydrostatic and aerostatic bearings to be used for Yates bearings. Stiffness values of Yates bearings are correspondingly reduced by similar factors as indicated above for load.

There is loss of load support when radial and axial loads are applied in combination, as evident in Figure 11.7. Values of $\overline{T}$ and $\overline{W}$ are given for different combinations of axial eccentricity ratio $\varepsilon_a = 1 - h_a/h_{ao}$ and radial eccentricity ratio $\varepsilon_r = e/h_{jo}$.

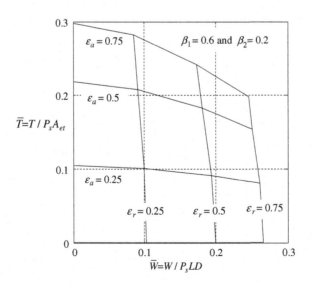

Figure 11.7: Typical Combined Axial and Radial Load Support for $L/D = 1$.

The following design procedure is based on combined radial and axial loads where the two eccentricity ratios are each 0.5. The design procedure is reasonably conservative but takes a Yates bearing closer to the absolute maximum than applying an eccentricity ratio of 0.5 for separate journal and thrust bearings. It is therefore prudent to ensure load values are not exceeded.

The effects of journal rotation are beneficial for radial loads and at optimum conditions they result in an increased radial load support of approximately 20%.

Figure 11.8 shows combined radial and thrust loads that may be applied with 0.5 eccentricity ratio in both the axial and radial directions. The figure is based on a journal bearing with four recesses and a 30° inter-recess land corresponding to a physical length $b = D/6$. For a six-recess journal bearing a 20° inter-recess land is recommended corresponding to a length $b = D/9$. For slot-entry bearings, the inter-slot separation is greatly reduced, as described in Chapter 10.

Journal bearing hydrostatic load support is increased with small values of a/L and a low thrust bearing pressure ratio β_2. For aerostatic journal plain bearings, $a/L = 0.25$ is usually

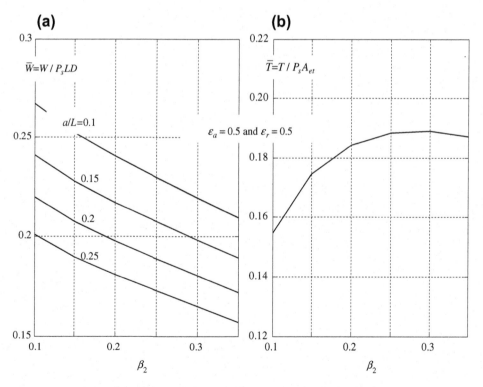

Figure 11.8: Yates Bearing: (a) Radial Load Support; (b) Thrust Load Support for $L/D = 1$.

appropriate. Axial load support increases as β_2 increases up to 0.3 approximately working within the recommended limits on eccentricity ratio. The chart assumes that the concentric pressure ratio of the journal bearing is set to $\beta_1 = (1 + \beta_2)/2$.

Effective Area of the Thrust Bearing

The effective area of the thrust bearing is the sum of the two sides:

$$A_{et} = A_{t1} + A_{t2} = 2A_t \tag{11.8}$$

This is in accordance with the definitions employed for opposed-pad bearings in Chapter 7. The area factor of a single circular pad with a central recess is

$$\overline{A}_t = \frac{A_t}{A} = \left[\frac{1 - D_i^2/D_o^2}{2\log_e(D_o/D_i)} \right] \tag{11.9}$$

where $A = \pi D_o^2/4$ is the area of one side before subtraction of the central shaft area.

Values of $\overline{A}_t$ for a single circular pad with a central recess are given in Figure 11.9. In the design procedure, the central shaft area is subtracted.

Bearing Film Stiffness

Average radial and axial stiffness values are estimated from loads supported at eccentricity ratio of 0.5. Average stiffness values are

$$\begin{aligned} \lambda_r &= 2W_{0.5}/h_j \\ \lambda_a &= 2T_{0.5}/h_j \end{aligned} \tag{11.10}$$

Alternatively, concentric radial stiffness factors $\overline{\lambda}_j$ for conventional hydrostatic journal bearings with various types of flow restrictors are given in Table 11.1. These expressions illustrate how various parameters affect stiffness. Values of $\overline{\lambda}_j$ may also be read from Figure 9.9. For a hydrostatic Yates bearings, the radial stiffness factor $\overline{\lambda}_r$ is reduced by a factor $1 - \beta_2$ so that

$$\overline{\lambda}_r = \overline{\lambda}_j(1 - \beta_2) \tag{11.11}$$

where

$$\lambda_r = \frac{P_s L D}{h_j} \overline{\lambda}_r$$

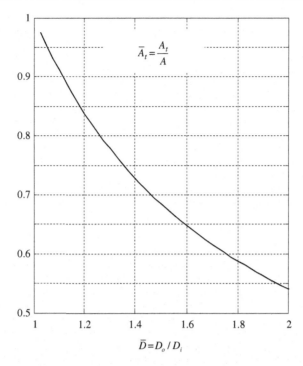

$$\bar{A}_t = \frac{A_t}{A}$$

$$\bar{D} = D_o / D_i$$

Figure 11.9: Yates Bearing: Area Factor for a Single Thrust Pad.

Table 11.1: Concentric radial stiffness factors $\bar{\lambda}_j$ for recessed journals

n	Capillary	Orifice	Constant Flow
4	$\dfrac{0.955K_{bs}}{1+0.5\gamma}$	$\dfrac{1.91K_{bs}}{1.5+\gamma}$	$\dfrac{1.91K_{bs}}{1+\gamma}$
5	$\dfrac{1.03K_{bs}}{1+0.345\gamma}$	$\dfrac{2.06K_{bs}}{1.5+0.69\gamma}$	$\dfrac{2.06K_{bs}}{1+0.69\gamma}$
6	$\dfrac{1.075K_{bs}}{1+0.25\gamma}$	$\dfrac{2.15K_{bs}}{1.5+0.5\gamma}$	$\dfrac{2.15K_{bs}}{1+0.5\gamma}$
	where $K_{bs} = (1 - a/L)$ for nongrooved journals		

Axial stiffness factors for a hydrostatic Yates bearing are given in Figure 11.10, where

$$\lambda_a = \frac{P_s A_{et}}{h_a} \bar{\lambda}_a \tag{11.12}$$

Friction Power

Bearing clearance and lubricant viscosity of hydrostatic bearings for high speeds should be adjusted to minimize total power and bring power ratio K into an acceptable range, as described

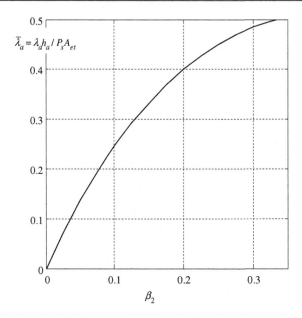

Figure 11.10: Stiffness Factors for a Yates Thrust Bearing.

in Chapter 3. The procedure below optimizes the bearing and acts to safeguard against risks associated with temperature rise and cavitation.

Friction power, H_f, is estimated from

$$H_f = \frac{\eta N^2 D^4}{2h_j} \cdot \overline{F} \tag{11.13}$$

The friction power factor is

$$\overline{F} = 2\pi^3 \left(\frac{L}{D}\overline{A}_{fj} + \frac{h_j}{h_a}\overline{A}_{ft} \right) \tag{11.14}$$

Friction factors for the journal bearing and the thrust bearing are related to the bearing dimensions by $\overline{A}_{fj} = A_{fj}/\pi DL$ and $\overline{A}_{ft} = A_{ft}/\pi D^2$. The friction factors are given in Figures 11.11 and 11.12.

11.4 Hydrostatic Design Procedure

1. Specify: maximum radial load, W.
2. Specify: maximum thrust load, T.
3. Specify: maximum operating speed, N.

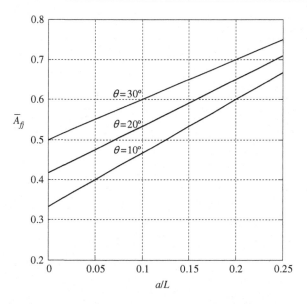

Figure 11.11: Friction Factor for a Yates Journal Bearing Where *a/L* is Land-Width Ratio and *θ* is Inter-Recess Land Arc.

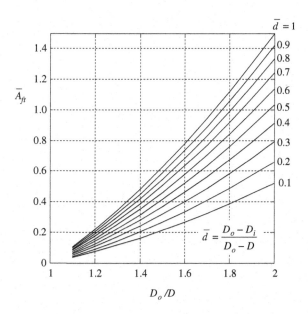

Figure 11.12: Friction Factor for a Yates Thrust Bearing.

4. Specify: supply pressure, P_s. If the bearing size is limited, supply pressure will be determined by the load values using Figure 11.8.

5. Specify: thrust pressure ratio, β_2. The practical range of β_2 values is limited to the range between 0.1 and 0.33. Suggestion: $\beta_2 = 0.2$.

6. Specify: journal pressure ratio β_1. Suggestion: $\beta_1 = (1 + \beta_2)/2$.
7. Specify: land-width ratio a/L for the journal. Suggestion: $a/L = 0.25$. For high speeds, this will sometimes result in viscosity lower than desired. Viscosity can be increased by reducing land-width ratio to $a/L = 0.1$ or if necessary by increasing journal clearance.
8. Read: $\overline{W} = W/P_sLD$ from Figure 11.8.
9. Calculate: minimum shaft diameter from $D = \sqrt{W/P_s\overline{W}}$.
10. Calculate: bearing length from $L = W/P_sD\overline{W}$. Suggestion: $L/D = 1$. Note: L/D should be greater than 0.5 and less than 1.5.
11. Specify: number of recesses or entry ports in the journal bearing. Minimum is $n = 4$ for reasonable load capacity and stiffness. Bearings of short L/D ratio should have at least five or six recesses. Two rows of 12 slots per row are recommended for plain slot-entry bearings. Minimum is eight slots per row.
12. Specify: inter-recess land width b or the inter-recess arc angle θ for recessed bearings. Suggestion: $\theta = 30°$ for $n = 4$ or $\theta = 20°$ for $n = 6$. Corresponding values of land width are $b = \pi D/12$ and $b = \pi D/18$.
13. Calculate: flow shape factor for journal from $\overline{B}_j = \pi D/6a$.
14. Specify: journal bearing radial film thickness h_j.
15. Specify: thrust bearing axial film thickness h_a.
16. Calculate: thrust pad shape factor required to achieve required pressure ratios from

$$\overline{B}_t = \frac{\overline{B}_j}{2}\left(\frac{\beta_1}{\beta_2} - 1\right)\left(\frac{h_j}{h_a}\right)^3$$

17. Read the diameter ratio for the thrust bearing $\overline{D}$ from Figure 11.6, using $\overline{B}_t$. Diameter ratio may also be estimated from $\overline{D} = 1 + (2\pi/(12\overline{B}_t - \pi))$.
18. Read: $\overline{T}$ from Figure 11.8.
19. Calculate: effective thrust bearing area from $A_{et} = T/P_s\overline{T}$.
20. Read: $\overline{A}_t$ from Figure 11.9.
21. Calculate: D_o. The outer diameter D_o must be larger than either $D_o = \overline{D}\cdot D$ or
    $D_o = \sqrt{(2A_{et}/\pi\overline{A}_t) + (D^2/\overline{A}_t)}$.
22. Calculate: $D_i = D_o/\overline{D}$. D_i must be greater than D. If not, increase D_i and increase D_o in proportion. Alternatively, P_s may be increased and D reduced from step 9.
23. Calculate: circumferential flow factor for recessed bearing $\gamma = [na(L - a)]/\pi Db$.
24. Calculate: $\overline{\lambda}_j$ from Table 11.1.
25. Calculate: radial stiffness factor from $\overline{\lambda}_r = \overline{\lambda}_j(1 - \beta_2)$.
26. Calculate: radial stiffness from $\lambda_r = (P_sLD/h_j)\cdot\overline{\lambda}_r$.
27. Read: axial stiffness factor $\overline{\lambda}_a$ from Figure 11.10.
28. Calculate: axial stiffness from $\lambda_a = (P_sA_{et}/h_a)\cdot\overline{\lambda}_a$.
29. Read: journal friction factor $\overline{A}_{fj}$ from Figure 11.11.
30. Read: thrust pad friction factor $\overline{A}_{ft}$ from Figure 11.12 or calculate from

$$\bar{A}_{ft} = \frac{D_o^2 - D_i^2}{2D^2} - \frac{3}{8}\frac{D_i^2 - D^2}{D^2}$$

31. Calculate: friction power factor from

$$\bar{F} = 2\pi^3 \left(\frac{L}{D}\bar{A}_{f1} + \frac{h_j}{h_a}\bar{A}_{f2} \right)$$

32. Calculate: flow factor from $\bar{Q} = 2\beta_2 \cdot \bar{B}_t$ taking $\bar{B}_t$ from step 16.
33. Calculate: optimum speed parameter from

$$S_{ho} = \sqrt{\frac{\bar{Q}}{8\bar{F}} \cdot \left(\frac{h_a}{h_j} \right)^3}$$

34. Calculate: optimum viscosity from $\eta = (S_{ho}P_s/N)(2h_j/D)^2$. (i) If η is too low, reduce a/L to not less than $200h_j$ or increase h_j. Repeat calculations from step 7. (ii) If η is too high, specify an acceptable value.
35. Calculate: flow from

$$q = 2\cdot\frac{P_s h_a^3}{\eta}\beta_2\bar{B}_t$$

36. Calculate: pumping power from $H_p = P_s q$.
37. Calculate: friction power from

$$H_f = \frac{\eta N^2 D^4}{2h_j}\cdot\bar{F}$$

38. Design control restrictors using the design procedures in Chapter 5. Flow through one restrictor is total flow divided by the number of restrictors.

■ Example 11.1 A Recessed Hydrostatic Yates Bearing Using Specified Values

Each step relates to the same numbered step in the above design procedure:
1. $W = 2224$ N (500 lbf).
2. $T = 1334$ N (300 lbf).
3. $N = 20$ rev/s.

4. $P_s = 2.068$ MN/m^2 (300 lbf/in^2).

5. $\beta_2 = 0.2$.

6. $\beta_1 = 0.6$.

7. $a/L = 0.25$.

8. $\overline{W} = 0.18$.

9. $D = \sqrt{2224/(2.068 \times 10^6 \times 0.18)} \cong 77.3$ mm (3.043 in).

10. $L = 77.3$ mm (3.043 in).

11. $n = 4$.

12. $b = (30°/360°) \times \pi \times 77.3 = 20$ mm (0.79 in).

13. $\overline{B}_j = (\pi \times 77.3)/(6 \times 0.25 \times 77.3) = 2.094$.

14. $h_j = 25.4$ µm (0.001 in).

15. $h_a = 25.4$ µm (0.001 in).

16. $\overline{B}_t = (2.094/2)(0.6/0.2 - 1)(25.4/25.4)^3 = 2.094$.

17. $\overline{D} = 1.3$.

18. $\overline{T} = 0.185$.

19. $A_{et} = 1334/(2.064 \times 10^6 \times 0.185) = 0.003494$ m^2 (5.416 in^2).

20. $\overline{A}_t = 0.775$.

21. $D_o = \sqrt{(2 \times 0.003494)/(\pi \times 0.775) + (0.0773^2/0.775)} = 0.1028$ m and
     $D_o(\text{min}) = 1.3 \times 0.0773 = 0.1005$ m. Say $D_o = 0.104$ m (4.1 in). This adjustment
     increases A_{et} slightly and allows room for a small thrust recess.

22. $D_i = 0.104/1.3 = 0.080$ m (3.15 in).

23. $\gamma = [4 \times 0.25 \times 77.3 \times (1 - 0.25) \times 77.3]/(\pi \times 77.3 \times 20) = 0.923$.

24. $\overline{\lambda}_j = [0.955 \times (1 - 0.25)]/(1 + 0.5 \times 0.923) = 0.49$.

25. $\overline{\lambda}_r = 0.49 \times (1 - 0.2) = 0.392$.

26. $\lambda_r = (2.068 \times 10^6 \times 0.0773 \times 0.0773)/(25.4 \times 10^6) \times 0.392 = 191$ MN/m
     (1.09×10^6 lbf/in).

27. $\overline{\lambda}_a = 0.4$ based on the effective area of two pads.

28. $\lambda_a = (2.068 \times 10^6 \times 0.003494)/(25.4 \times 10^6) \times 0.4 = 114$ MN/m (0.65×10^6
     lbf/in).

29. $\overline{A}_{fj} = 0.74$.

30. $\overline{A}_{ft} = (0.1005^2 - 0.0773^2)/(2 \times 0.0773^2) - {}^3/_8 \times (0.080^2 - 0.0773^2)/$
     $0.0773^2 = 0.345 - 0.0267 = 0.319$.

31. $\overline{F} = 2\pi^3[1 \times 0.74 + (25.4/25.4) \times 0.319] = 65.7$.

32. $\overline{Q} = 2 \times 0.2 \times 2.094 = 0.838$.

33. $S_{ho} = \sqrt{[0.838/(8 \times 65.7)] \times (25.4/25.4)^3} = 0.04$.

34. $\eta = [(0.04 \times 2.068 \times 10^6)/20](2 \times 25.4 \times 10^{-6}/0.0773)^2 = 0.0018$ N s/m^2 (1.8 cP
     or 0.26×10^{-6} reyns).

35. $q = [(2.068 \times 10^6 \times 25.4^3 \times 10^{-18})/0.0018] \times 0.838 = 15.8 \times 10^{-6}$ m^3/s (0.95
     l/min).

36. $H_p = 2.068 \times 10^6 \times 15.8 \times 10^{-6} = 32.7$ W.
37. $H_f = [(0.0018 \times 20^2 \times 0.0773^4)/(2 \times 25.4 \times 10^{-6})] \times 65.7 = 33.2$ W.

For an optimized bearing, the value of H_p should equal the value of H_f and this condition forms a useful check in calculations.

■

11.5 Aerostatic Yates Bearings

Yates bearings can be designed using the above data and a simpler procedure. Friction power is usually negligible except at very high speeds and is often ignored. Recesses are avoided to minimize the volume of compressible gas within the bearing. A large recess volume reduces dynamic performance and risks pneumatic instability, particularly with orifice restrictors. For best performance with $0.5 < L/D < 1.5$, a two-row, 12-slot plain bearing is recommended. The geometry of plain journal bearings was described in the previous chapter. A suitable geometry for a thrust bearing geometry for an aerostatic Yates bearing is shown in Figure 11.13.

Aerostatic loads. Bearing loads supported by aerostatic bearings are similar to hydrostatic bearings. The limits suggested should not be exceeded.

Aerostatic flow. Flow through a concentric aerostatic Yates bearing based on the thrust bearing is

$$q = \left(\frac{p_2^2 - p_a^2}{2p_a} \right) \frac{h_a^3}{\eta} \cdot \bar{B}_t$$

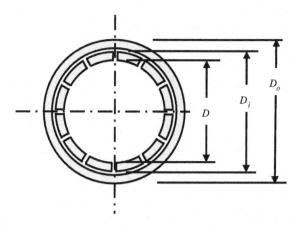

Figure 11.13: Shallow Grooves Minimize Recess Volume in an Aerostatic Pad.

where $\bar{B}_t$ is given by equation (11.4) or alternatively, based on the journal bearing,

$$q = \left(\frac{p_1^2 - p_2^2}{2p_a}\right)\frac{h_j^3}{\eta}\cdot\bar{B}_j$$

where $\bar{B}_j. = \pi D/6a$. The absolute pressure p_1 stands for the concentric inlet pressure from the slot restrictors to the journal, and p_2 denotes the concentric pressure between the journal lands and the thrust lands. Flow from the above expression is in terms of free air at ambient pressure p_a. Flow through the two thrust pads equals flow through the journal bearing, so that

$$\bar{B}_t = \frac{\bar{B}_j}{2}\left(\frac{p_1^2 - p_2^2}{p_2^2 - p_a^2}\right)\left(\frac{h_j}{h_a}\right)^3 \tag{11.15}$$

11.6 Aerostatic Design Procedure

1. Specify: maximum radial load, W.
2. Specify: maximum thrust load, T.
3. Specify: maximum operating speed, N.
4. Specify: gauge supply pressure, $P_s = p_s - p_a$. If the bearing size is limited, the supply pressure will be determined by the load values using Figure 11.8.
5. Specify: thrust pressure ratio, K_{go2}. A practical range is $0.1 < K_{go2} < 0.33$. Suggestion: $K_{go2} = 0.2$.
6. Specify: journal pressure ratio K_{go1}. Suggestion: $K_{go1} = (1 + K_{go2})/2$.
7. Calculate: absolute concentric pressures from $p_1 = K_{go1} \cdot P_s + p_a$ and $p_2 = K_{go2} \cdot P_s + p_a$.
8. Specify: land-width ratio a/L for the journal bearing. Suggestion: $a/L = 0.25$.
9. Read: $\bar{W} = W/P_sLD$ from Figure 11.8.
10. Calculate: shaft diameter from $D = \sqrt{W/P_s\bar{W}}$.
11. Calculate: bearing length from $L = W/P_sD\bar{W}$. Suggestion: $L/D = 1$ or $0.5 < L/D < 1.5$.
12. Calculate: land width from $a = L(a/L)$.
13. Specify: number of entry ports. Two rows of 12 slots per row are recommended.
14. Read: $\bar{T}$ from Figure 11.8.
15. Calculate: minimum thrust area from $A_{et} = T/P_s\bar{T}$.
16. Estimate: thrust bearing diameter ratio from $D_o = \sqrt{2A_{et} + D^2}$.
17. Check:

$$A_{et} = \frac{\pi}{2}\left[\frac{D_o^2 - D^2}{2\ln(D_o/D)} - D^2\right] > \text{Minimum } A_{et}$$

from (11.14). If not, increase D_o or increase P_s.
18. Calculate: shape factor for journal flow from $\bar{B}_j = \pi D/6a$.

19. Read: shape factor for thrust pad flow from Figure 11.6 or calculate from

$$\overline{B}_t = \frac{\pi}{6 \ln (D_o/D)}$$

20. Calculate: film thickness ratio:

$$\frac{h_a}{h_j} = \left[\frac{\overline{B}_j}{2\overline{B}_t} \left(\frac{p_1^2 - p_2^2}{p_2^2 - p_a^2} \right) \right]^{1/3}$$

21. Specify: radial film thickness h_j for the journal bearing.
22. Specify: axial film thickness h_a for the thrust bearing.
23. Estimate: average radial stiffness λ_r from $\lambda_r = 2W/h_j$.
24. Estimate: average axial stiffness from $\lambda_a = 2T/h_a$.
25. Calculate: free air flow from

$$q_a = 2 \cdot \frac{P_s \overline{B}_t h_a^3}{\eta} K_{go2} \cdot \frac{p_2 + p_a}{2p_a}$$

26. Calculate pumping power dissipation from $H_p = p_a q_a \ln(p_s/p_a)$.
27. Design flow restrictors using the design procedures at the end of the book. *Warning*: Flow through a restrictor is total flow divided by the number of restrictors.

■ *Example 11.2 An Aerostatic Double-Row Yates Bearing Using Specified Values*

Each step relates to the same numbered step in the above design procedure:
1. $W = 2224$ N (500 lbf), radial load.
2. $T = 1334$ N (300 lbf), axial load.
3. $N = 20$ rev/s, journal speed.
4. $P_s = 0.4$ MN/m^2 (58 lbf/in^2), gauge supply pressure.
5. $K_{go2} = 0.2$, gauge pressure ratio for thrust bearing $= P_2/P_s$.
6. $K_{go1} = 0.6$, gauge pressure ratio for journal bearing $= P_1/P_s$.
7. $p_1 = (0.6 \times 0.4 + 0.101) \times 10^6 \equiv 0.341$ MN/m^2 absolute pressure.
    $p_2 = (0.2 \times 0.4 + 0.101) \times 10^6 \equiv 0.181$ MN/m^2 absolute pressure.
    $P_2 = K_{go2} P_s = 0.2 \times 0.4 = 0.08$ MN/m^2 gauge pressure.
8. $a/L = 0.25$, land-width ratio for journal bearing.
9. $\overline{W} = 0.175$, load factor for journal bearing.

10. $D = \sqrt{2224/(0.4 \times 10^6 \times 0.175)} \equiv 178$ mm (7.0 in), journal diameter.
11. $L = 178$ mm (7.0 in), journal bearing length.

12. $a = 0.25 \times 178 = 44.5$ mm, journal bearing land width.

13. $n = 12$, two rows of 12 slots per row.

14. $\overline{T} = 0.185$, load factor for thrust bearing.

15.   Minimum $A_{et} = 1334/(0.4 \times 10^6 \times 0.185) = 0.01802$ m^2 (27.93 in^2), total eff. thrust area.

16. $D_o = \sqrt{2A_{et} + D^2} = \sqrt{2 \times 0.01802 + 0.178^2} = 0.26$ m (10.24 in).

17. $A_{et} = \dfrac{\pi}{2}\left[\dfrac{260^2 - 178^2}{2\ln(260/178)} - 178^2\right] \times 10^{-6} = 0.0247$ m^2 and is greater than minimum.

18. $\overline{B}_j = (\pi \times 178)/(6 \times 44.5) = 2.094$, journal bearing shape factor for flow.

19. $\overline{B}_t = \pi/[6\ln(260/178)] = 1.38$, shape factor for thrust pad flow.

20. $h_a/h_j = [(2.094/2 \times 1.38)(0.341^2 - 0.181^2)/(0.181^2 - 0.101^2)]^{1/3} = 1.41$, film thickness ratio.

21. $h_j = 20$ μm (0.00079 in), journal bearing film thickness.

22. $h_a = 20 \times 1.41 = 28.2$ μm (0.0011 in), thrust pad film thicknes.

23. $\lambda_r = 2 \times 2224/20 \times 10^{-6} = 222$ MN/m, average radial stiffness.

24. $\lambda_a = 2 \times 1334/28.2 \times 10^{-6} = 94.6$ MN/m, average axial stiffness.

25. $q_a = [2 \times (0.4 \times 10^6 \times 1.38 \times 28.2^3 \times 10^{-18})/(18.2 \times 10^{-6})] \times 0.2 \times (0.181 + 0.101)/(2 \times 0.101) \equiv 0.38$ l/s free air.

26. $H_p = 0.101 \times 10^6 \times 0.00038 \times \ln(0.4/0.101) = 52.8$ W.

27.   Design flow restrictors using the design procedures at the end of the book. *Warning*: Flow through a restrictor is total flow divided by the number of restrictors.

■

References

Lund, J. W. (1963). Static stiffness and dynamic angular stiffness of the combined hydrostatic journal–thrust bearing. Mechanical Technology Inc., Report Number MTI 63 TR45.

Wearing, R. S., O'Donoghue, J. P., & Rowe, W. B. (1970). Design of combined journal–thrust hydrostatic bearings (the Yates bearing). *Machinery and Production Engineering, 117*, 301–308, (3014; 19 August).

Yates, H. G. (1950). *Combined journal and thrust bearings.* UK Patent 639, 293, June.

CHAPTER 12

Conical Journal Bearings

Summary of Key Design Formulae

$$q = \frac{P_s h_o^3}{\eta} \frac{D}{a} \cdot QF$$ Hydrostatic flow

$$q = \frac{P_s h_o^3}{\eta} \frac{D}{a} \cdot QF \cdot \frac{p_{\text{ro}} + p_a}{2p_a}$$ Aerostatic flow

$$W = P_s LD \cdot \overline{W}$$ Radial load support

$$T = P_s D^2 \cdot \overline{T}$$ Axial load support

$$S_h = \frac{\eta N}{P_s} \left(\frac{D}{2h_o} \right)^2$$ Speed parameter

12.1 Application

Conical hydrostatic and aerostatic journal bearings have two advantages. By replacing two bearings with one there is economy of flow, economy of power, and fewer parts. Secondly, clearance is adjustable on assembly. Against these advantages are increased requirements for accurate machining and accurate assembly of conical surfaces. Conical spindle bearings were applied by Rowe (1967) for a precision grinding spindle and early design procedures were developed by Aston et al. (1970, 1971).

Geometry of a recessed conical bearing is shown in Figure 12.1 and a recessed bearing is compared with a plain bearing in Figure 12.2. The plain configuration is suitable for hydrostatic, hybrid, or aerostatic application. Plain bearings will usually have slot restrictors, although capillary restrictors or orifice restrictors may be employed preferably with slot-shaped entry pockets.

Recesses are avoided for aerostatic bearings to minimize the risk of pneumatic hammer instability. This is because large volumes of compressed gas within the bearing are associated with poor dynamic performance.

Hydrostatic, Aerostatic and Hybrid Bearing Design.
DOI: 10.1016/B978-0-12-396994-1.00012-7

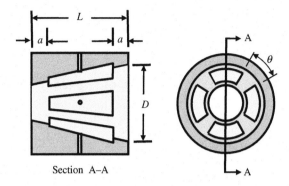

Figure 12.1: Geometry of the Recessed Conical Journal Bearing.

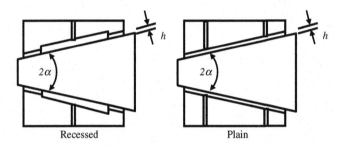

Figure 12.2: Recessed and Plain Conical Bearings.

Recesses are avoided for high-speed hybrid hydrostatic bearings too, since plain lands support additional hydrodynamic load.

Data charts are presented below for recessed and plain conical bearings. Recessed bearing data are for four recesses, although five or six recesses have slightly superior load and stiffness performance. Plain slot-entry bearings usually have two rows of 12 slots per row. Such bearings give better load performance than a four-recess bearing but create slightly more friction drag. Hydrostatic data for four-recess bearings can also be used as a guide for hydrostatic and aerostatic plain bearing performance. Application to hydrostatic and aerostatic design is illustrated by examples later in the chapter. Aerostatic design is simpler in that friction power is almost always negligible unless very high speeds are employed together with small clearances.

Each recess is supplied from a constant gauge supply pressure P_s by way of an individual flow control device. A capillary or slot restrictor is needed for each recess or slot entry.

Alternative configurations for conical spindle bearings are illustrated in Figure 12.3. Spindle bearings employ opposed complementary-cone arrangements, as shown. Better load-carrying

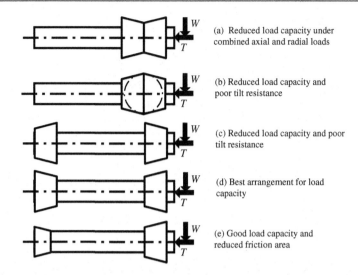

(a) Reduced load capacity under combined axial and radial loads

(b) Reduced load capacity and poor tilt resistance

(c) Reduced load capacity and poor tilt resistance

(d) Best arrangement for load capacity

(e) Good load capacity and reduced friction area

Figure 12.3: A Comparison of Five Conical Bearing Arrangements.

capacity is achieved if the cones are well separated rather than closely spaced. Wide spacing allows either central application of radial loads or overhung application of radial loads, as illustrated in Figure 12.3. For many machines, overhung application is more usual.

The effect of close spacing, as in Figure 12.3a, may be appreciated by considering the resulting clearances. An axial thrust load displaces the shaft axially so that clearance in one cone is reduced. A centrally applied radial displacement of the shaft reduces the minimum film thickness of each bearing. However, maximum radial displacement is governed by the small-clearance cone and at this condition the small-clearance cone carries a large radial load while the large-clearance cone carries little radial load. The advantage of the widely spaced arrangement is that just a slight tilt allows the large-clearance cone to carry increased load, hence the total radial load supported is almost doubled.

The face-to-face configuration (Figure 12.3b) has negligible aligning capability. The pads are disposed approximately around a circle, as shown dotted in the figure, so that another bearing is required to maintain alignment of the bearing clearances. Alignment is simplified with the widely spaced arrangements in Figure 12.3c and d.

For overhung loads applied at one end, as shown in Figure 12.3, the reaction at the bearing nearest the load is greater than the applied load whilst the reaction at the far bearing is reduced. The calculation is illustrated in Figure 12.4. It helps if the near bearing is larger than the far bearing, as in Figure 12.3e. Power is reduced and assembly is more convenient. Axial projected areas should be approximately equal.

Data are given for the maximum radial load supported on one bearing while the maximum thrust load is given for the opposed pair acting together, as in Figure 12.4. The designer

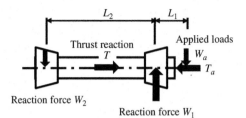

Equilibrium of applied and reaction forces requires that:
$$W_1 = W_a \cdot (L_1 + L_2) / L_1$$
$$W_2 = W_a \cdot L_1 / L_2$$
$$T = T_a$$

Figure 12.4: Bearing Forces for Loads Applied at One End.

must first specify the maximum radial and thrust reaction loads. Maximum load W_1 from Figure 12.4 is used as the radial load W in the design procedure. Specified radial and thrust loads must not be exceeded.

Flow is greater from the large end of a plain conical bearing than from the small end. Ideally, this requires the restrictors to have a larger flow resistance at the small end of the cone.

12.2 Basic Parameters

Main parameters are listed in the "key design formulae" above. Data are given in charts.

Geometry

Basic geometry is illustrated in Figures 12.1 and 12.2:

D	Bearing diameter at large end of cone
L	Axial bearing length
a	Axial land width
b	Inter-recess land width; varies with position along length
$a\sec\alpha$	Land width measured along cone surface
θ	Angle of inter-recess land
h_o	Film thickness normal to bearing surface

Lengths along the conical surface are increased by $\sec\alpha$. Diameter D applies to the large end of the cone and is not the average diameter. The inter-recess land, b, is specified by the arc θ.

Displacement and Minimum Film Thickness

Radial eccentricity ratio is $\varepsilon = e_r/h_{ro}$. Axial film thickness is $\overline{X} = h_a/h_{ao}$. Minimum film thickness is $h_{min} = (1 - \varepsilon)\overline{X} \cdot h_o$. When axial and radial loads are employed together, it is advisable to use conservative displacement values, particularly for the cone having reduced film thickness. Axial eccentricity ratio of 0.3 for a complementary-cone arrangement gives film thickness values of $\overline{X} = 0.7$ and 1.3. Combining $\varepsilon = 0.5$ and $\overline{X} = 0.7$ reduces minimum film thickness to $0.35h_o$. These maximum displacements should not be exceeded.

Pressure Ratio and Restrictor Design

The symbol for concentric pressure ratio is β for hydrostatic bearings and K_{go} for aerostatic bearings. Pressure ratio is defined as $\beta = K_{go} = P_{ro}/P_s = (p_{ro} - p_a)/(p_s - p_a)$. Gauge supply pressure is P_s whereas absolute supply pressure is p_s. Similarly, gauge entry pressure is P_{ro} and absolute entry pressure is p_{ro}. Absolute ambient pressure is p_a. Absolute pressures are required for calculation of aerostatic flows.

Restrictors are designed to achieve a reduced range $0.4 < \beta < 0.5$. Exceeding the range $0.4 < \beta < 0.5$ leads to unacceptable pressure ratios after applying an axial displacement. Axial film thickness $\overline{X} = 0.75$ modifies concentric pressure ratio from $\beta = 0.5$ to effective values $\beta_e = 0.7$ and $\beta_e = 0.34$. This combination allows reasonable load support in both cones and is judged acceptable. Load range is slightly improved by reducing pressure ratio to $\beta = 0.4$ with axial film thickness $\overline{X} = 0.7$. For hydrostatic bearings, it is suggested that $\beta = 0.4$ for complementary cones. For aerostatic bearings, it is suggested that $K_{go} = 0.5$.

After a bearing has been designed, flow per restrictor is used to design restrictors, as in Chapter 5. Flow is greater at the large end of the bearing and smaller at the small end. This suggests that restrictors should be designed appropriately to match the different flows at each end.

Data charts given below are based on capillary or slot control. If orifice control is employed, slightly more conservative values of load capacity should be assumed. In all cases, it is recommended that larger bearings than otherwise are employed to allow for possible overloads.

Flow from Each End

Flow factors QF are given below for single-cone bearings. Flow factor $QF1$ from the large end of a cone is larger than flow factor $QF2$ from the small end, where $QF = QF1 + QF2$. As mentioned above, difference in flows at the two ends affects the design of restrictors for the small end of plain conical bearings whereas the large end is hardly affected.

Aerostatic free air flow is larger than hydrostatic flow by a factor $(p_r + p_a)/2p_a$.

12.3 Single-Cone Bearings

Load Data

Load data for thin-land single-cone hydrostatic bearings are given in Figure 12.5 for $L/D = 0.5$ and in Figure 12.6 for $L/D = 1.0$ with pressure ratio $\beta = 0.5$. The land width $a/L = 0.1$ and the inter-recess angle $\theta = 30°$. Data are presented for three values of semi-cone angle α. The practical range for $L/D = 0.5$ is shown from $\alpha = 10°$ to $\alpha = 30°$. For $L/D = 1.0$, the practical range is shown as $\alpha = 10°$ to $\alpha = 20°$. For land width increased to $a/L = 0.25$, load values are reduced by 17%.

Large L/D ratios are only possible with very small cone angles due to the constraint imposed by the inner diameter of the bearing. The L/D values shown therefore cover the usual operating range. The figures show maximum radial loads for a radial eccentricity $\varepsilon = 0.5$ and axial film thickness values within the range $\overline{X} = 0.5$ to $\overline{X} = 2.0$.

When a maximum axial displacement is applied at the same time as a maximum radial displacement, minimum film thickness is reduced. For example, 0.5 axial displacement together with a 0.5 radial displacement results in $h_{min}/h_o = (1 - \varepsilon) \cdot \overline{X} = 0.25$. The combined radial and axial displacements are therefore absolute maximum values.

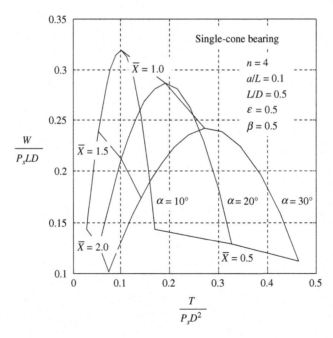

Figure 12.5: Design Data for Single-Cone Bearings, $L/D = 0.5$, $\beta = 0.5$.

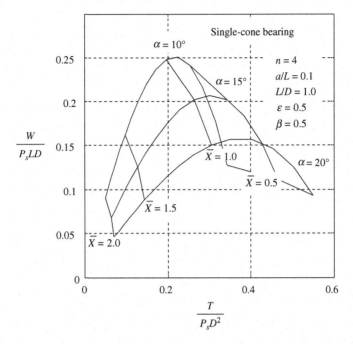

Figure 12.6: Design Data for Single-Cone Bearings, $L/D = 1.0$, $\beta = 0.5$.

The effect of reduced pressure ratio $\beta = 0.4$ is shown in Figures 12.7 and 12.8. A reduced pressure ratio improves radial load support as axial film thickness is reduced. Conversely, radial load support is reduced as axial film thickness is increased.

Flow Data

Flow data for single-cone bearings are given in Figure 12.9. Flow for plain bearings can be split into two terms. The flow factor for the large end is $QF1$ and for the small end is $QF2$, where $QF = QF1 + QF2$. This is illustrated for $L/D = 1$ where the difference is greatest. Whereas flow at the large end is hardly affected by cone angle, flow at the small end is greatly reduced.

12.4 Complementary-Cone Spindle Bearings

Flow and radial load data for single-cone bearings with suitable interpretation may also be used for complementary-cone arrangements. Axial thrust load for two cones acting in opposition is given by $T = T_1 - T_2$. Values T_1 and T_2 may be read from the charts for single-cone bearings.

The first decision, however, is a suitable combination of semi-cone angle α and length-to-diameter ratio L/D. Data for this purpose are provided in Figure 12.10 based on the ratio of the applied axial thrust load T to the applied radial load W. For larger cone angles it

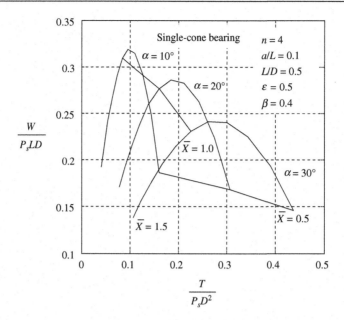

Figure 12.7: Design Data for Single-Cone Bearings, *L/D* = 0.5, *β* = 0.4.

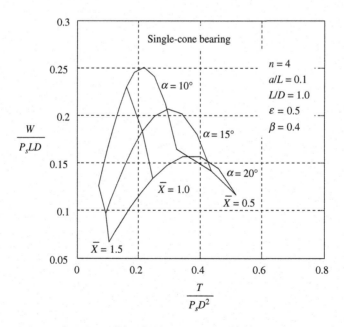

Figure 12.8: Design Data for Single-Cone Bearings, *L/D* = 1.0, *β* = 0.4.

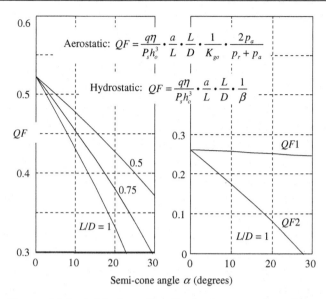

$$\text{Aerostatic: } QF = \frac{q\eta}{P_s h_o^3} \cdot \frac{a}{L} \cdot \frac{L}{D} \cdot \frac{1}{K_{go}} \cdot \frac{2p_a}{p_r + p_a}$$

$$\text{Hydrostatic: } QF = \frac{q\eta}{P_s h_o^3} \cdot \frac{a}{L} \cdot \frac{L}{D} \cdot \frac{1}{\beta}$$

Semi-cone angle α (degrees)

Figure 12.9: Guide Data for Single-Cone Bearing Flow.

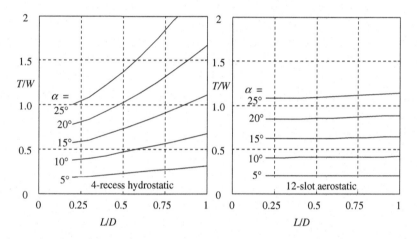

Figure 12.10: Selection of L/D for Conical Spindle Bearings.

becomes impracticable to accommodate L/D larger than $L/D = 1$. For $L/D = 1$, the maximum practicable value of semi-cone angle is $\alpha = 25°$.

Figures 12.11 and 12.12 give the resultant axial thrust for two complementary cones acting in opposition. The radial support load is for the cone where the axial film thickness is reduced and the radial load is maximal. For radial loads applied at one end, it is more appropriate to arrange that the bearing nearest the load is the one with reducing film thickness.

Figure 12.11 shows maximum radial and axial loads that may be applied in combination for a four-recess hydrostatic conical bearing. Short L/D bearings allow higher radial

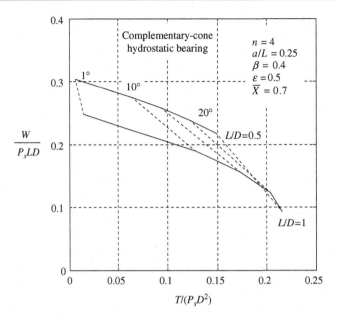

Figure 12.11: Load Data: Recessed Hydrostatic Spindle Bearings.

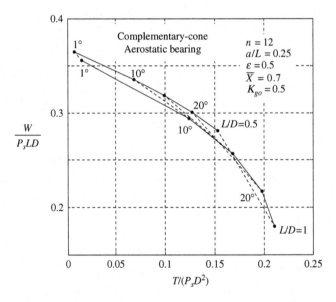

Figure 12.12: Load Data: Aerostatic Spindle Bearings.

loads per unit area than longer L/D bearings. With $L/D = 1$, larger thrust loads are possible but radial load capacity reduces as axial thrust loads are increased. A plain double-entry bearing supports higher load, as shown in Figure 12.12 computed for aerostatic supply.

Average stiffness values for $\varepsilon = 0.5$ and $\overline{X} = 0.7$ can be estimated as follows:

$$\lambda_r = 2W_{0.5} \cos \alpha / h_o$$
$$\lambda_a = 3.3T_{0.7} \sin \alpha / h_o \qquad (12.1)$$

12.5 High-Speed Design

At higher speeds, minimum power is achieved when pumping power H_p is equal to friction power H_f. The optimum lies in the range $1 < K < 3$, where $K = H_f/H_p$. The parameter S_h can be employed for selection of combinations of viscosity, supply pressure, and film thickness, where

$$S_h = \frac{\eta N}{P_s} \cdot \left(\frac{D}{2h_o} \right)^2$$

Values of S_{ho} are for $K = 1$, and are given in Figure 12.13. For other values of power ratio and pressure ratio, $S_h = S_{ho}\sqrt{K\beta}$.

12.6 Design Procedure for a Complementary-Cone Hydrostatic Bearing

1. Specify: maximum radial load to be carried by the main bearing, W.
2. Specify: maximum axial thrust, T.
3. Specify: operating speed in rev/s, N.
4. Specify: supply pressure, P_s.
5. Specify: pressure ratio, β. Suggest $\beta = 0.4$.
6. Select: combination of L/D and α from Figure 12.10 for T/W.
7. Specify: number of recesses n or slots per row n.
8. Specify: land-width ratio a/L and inter-recess angle θ. If possible make $a/L = 0.25$ and $\theta = 30°$.
9. Read: thrust factor $\overline{T}$ from Figure 12.11 or 12.12 as appropriate.
10. Calculate: D from $D = \sqrt{T/P_s\overline{T}}$.
11. Calculate: L from $L = D \times L/D$.
12. Read: radial load support factor, $\overline{W}$, from Figure 12.11.
13. Calculate: $W = P_sLD \cdot \overline{W}$ and compare with W required.
14. Specify: normal clearance h_o (shown in Figure 12.2).
15. Read: optimization parameter S_{ho} from Figure 12.13. If $N = 0$, specify a convenient value of viscosity and proceed to step 18.
16. Calculate: viscosity from $\eta = (S_{ho}P_s/N)(2h_o/D)^2$. Continue from step 18 unless the calculated value of η is too low, in which case specify the minimum acceptable value. If calculated η is too high, specify a convenient value and continue from step 18.

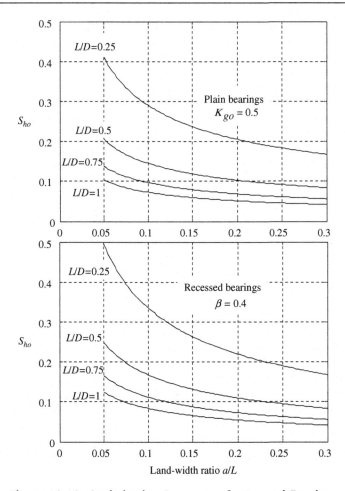

Figure 12.13: Optimization Parameter for Journal Bearings.

17. Calculate: new S_h for the minimum viscosity from $S_h = (\eta N/P_s)(D/2h_o)^2$.
18. Read: new a/L from Figure 12.13. Continue from step 18 unless $a < 50h_o$, in which case specify $a = 50h_o$.
19. Read: new S_{ho} for minimum land-width ratio from Figure 12.13.
20. Calculate: new clearance from $h_o = (D/2)\sqrt{\eta N/P_s S_{ho}}$.
21. Read: flow factor from Figure 12.9.
22. Calculate: flow for two bearing cones from

$$q = \frac{2P_s h_o^3}{\eta} \frac{D\,L}{L\,a} \cdot \beta \cdot QF$$

23. Calculate: radial stiffness of each bearing from $\lambda_r = (2W_{0.5}/h_o)\cos\alpha$.
24. Calculate: axial stiffness from $\lambda_a = (3.3T_{0.7}/h_o)\sin\alpha$.

25. Calculate: pumping power from $H_p = P_s q$.
26. Calculate: total power. If $N = 0$, $H_t = H_p$. If $N \neq 0$, $H_t = 2H_p$.
27. Calculate: maximum temperature rise of the lubricant for a single pass through the bearing from $\Delta T = H_t / c\rho q$, where c is the specific heat capacity of the lubricant and ρ is the density of the lubricant.
28. Calculate: flow per restrictor. For a four-recess complementary-cone bearing, $q_{restrictor} = q/8$. For a plain complementary-cone bearing with two rows of 12 slots per row, restrictor flow at large end, $q_{restrictor} = q/48$. For small end, the reduced flow per restrictor is $q_{restrictor} = 2(q/48) \cdot QF2/QF$ (see Figure 12.9).
29. Design restrictors: use the flow per restrictor and refer to Chapter 5.

■ Example 12.1 A Complementary-Cone Recessed Hydrostatic Bearing

1. $W = 2224$ N (500 lbf).
2. $T = 1334$ N (300 lbf).
3. $N = 20$ rev/s.
4. $P_s = 2.068$ MN/m^2 (300 lbf/in^2).
5. $\beta = 0.4$.
6. $L/D = 0.5$ and $\alpha = 10°$.
7. $n = 4$ recesses.
8. $a/L = 0.25$ and $\theta = 30°$.
9. $\overline{T} = 0.069$.
10. $D = \sqrt{1334/(2.068 \times 10^6 \times 0.069)} = 0.0967$ m (3.81 in).
11. $L = 0.0967 \times 0.5 = 0.0483$ m (1.9 in).
12. $\overline{W} = 0.27$.
13. $W = 2.068 \times 10^6 \times 0.0483 \times 0.0967 \times 0.27 = 2600$ N (586 lbf) compared with 2224 N minimum.
14. $h_o = 25.4$ μm (0.001 in).
15. $S_{ho} = 0.095$.
16. $\eta = [(0.0095 \times 2.068 \times 10^6)/20](50.8 \times 10^{-6}/0.0967)^2 = 0.0027$ N s/m^2 or 2.7 cP. At this stage it may be decided that this value is too low. For example, the minimum acceptable viscosity may be $\eta = 9$ cP.
17. New value of S_h:

$$\frac{0.009 \times 20}{2.068 \times 10^6}\left(\frac{0.0967}{50.8 \times 10^{-6}}\right)^2 = 0.315$$

18. $a/L = 0.1$.
19. $S_{ho} = 0.17$.
20. $h_o = (0.0967/2)\sqrt{(0.009 \times 20)/(2.068 \times 10^6 \times 0.17)} = 34.6 \times 10^{-6}$ m (0.0014 in).

21. Flow factor $= 0.48$.
22. $q = (2 \times 2.068 \times 10^6 \times 34.6^3 \times 10^{-18} \times 0.4 \times 0.48)/(0.009 \times 0.1 \times 0.5) = 0.000073$ m^3/s or 0.073 l/s.
23. $\lambda_r = [(2 \times 2600)/(34.6 \times 10^{-6})] \times 0.98 = 147$ MN/m $(0.84 \times 10^6$ lbf/in).
24. $\lambda_a = [(3.3 \times 1334)/(34.6 \times 10^{-6})] \times 0.174 = 22$ MN/m $(0.13 \times 10^6$ lbf/in).
25. $H_p = 2.068 \times 10^6 \times 0.00007 = 145$ W.
26. $H_t = 2 \times 145 = 290$ W.
27. If $c = 2120$ J/kg K and $\rho = 855$ kg/m^3, $\Delta T = 290/(2120 \times 855 \times 0.00007) = 2.3$ K.
28. $q_{restrictor} = 0.000073/8 = 0.000009125$ m^3/s or 0.009125 l/s.
29. Design restrictors using Appendix 5.A.

■

12.7 Design Procedure for a Complementary-Cone Aerostatic Bearing

1. Specify: maximum radial load to be carried by the main bearing, W.
2. Specify: maximum axial thrust, T.
3. Specify: operating speed in rev/s, N.
4. Specify: gauge supply pressure, $P_s = p_s - p_a$.
5. Specify: gauge pressure ratio, K_{go}. Suggest $K_{go} = 0.5$.
6. Calculate: absolute inlet pressure, $p_{ro} = p_a + K_{go}P_s$.
7. Select a suitable combination of L/D and α from Figure 12.10 for T/W.
8. Specify: number of slots per row n for a double-entry bearing. Suggest $n = 12$.
9. Specify: land-width ratio a/L. Suggest $a/L = 0.25$.
10. Read: thrust factor $\overline{T}$ from Figure 12.12.
11. Calculate: D from $D = \sqrt{T/P_s\overline{T}}$.
12. Calculate: L from $L = D \times L/D$.
13. Read: radial load factor $\overline{W}$ from Figure 12.12.
14. Calculate: $W = P_sLD \cdot \overline{W}$ and compare with required W.
15. Specify: normal bearing clearance h_o (shown in Figure 12.2).
16. Specify: viscosity. For air at 18 °C, $\eta = 18.3 \times 10^{-6}$ N s/m^2.
17. Read: flow factors QF and $QF2$ from Figure 12.9.
18. Calculate: free air flow for two cones from

$$q = \frac{2P_sh_o^3}{\eta}\cdot\frac{L}{a}\cdot\frac{D}{L}K_{go}\cdot\frac{p_r + p_a}{2p_a}QF$$

19. Estimate: average radial stiffness of each bearing from $\lambda_r = (2W_{0.5}/h_o)\cos\alpha$.
20. Estimate: average axial stiffness from $\lambda_a = (3.3T_{0.7}/h_o)\sin\alpha$.
21. Calculate: pumping power from $H_p = p_aq_a\ln(p_s/p_a)$.

22. Check (optional): friction power from $H_f = (\eta A_f U^2)/h_o$ with $A_f \approx \pi DL$ and $U = \pi DN$.
23. Calculate: total power $H_t = H_p + H_f$.
24. Calculate: flow per large-end restrictor from $q_{restrictor} = q/48$ for two double-row cones and $n = 12$. Flow per small-end restrictor from $q_{restrictor} = 2(q/48)(QF2/QF)$.
25. Design restrictors using the flow per restrictor and a procedure in Chapter 5.

■ *Example 12.2 A Complementary-Cone Double-Entry Aerostatic Bearing*

1. $W = 400$ N (90 lbf).
2. $T = 200$ N (45 lbf).
3. $N = 100$ rev/s.
4. $P_s = 0.5$ MN/m^2 (72.5 lbf/in^2).
5. $K_{go} = 0.5$.
6. $p_{ro} = 0.101 + 0.5 \times 0.5 = 0.351$ MN/m^2 (50.91 lbf/in^2) for $p_a = 0.101$ MN/m^2.
7. $L/D = 1.0$ and $\alpha = 10°$. Note: $12°$ would be suitable.
8. $n = 12$ slots per row with double-entry complementary cones.
9. $a/L = 0.25$.
10. $\overline{T} = 0.125$.
11. $D = \sqrt{200/(0.5 \times 10^6 \times 0.125)} = 0.0566$ m (2.23 in).
12. $L = 0.0566$ m (2.23 in).
13. $\overline{W} = 0.29$.
14. $W = 0.5 \times 10^6 \times 0.0566 \times 0.0566 \times 0.29 = 464$ N (104 lbf).
15. $h_o = 20$ μm (0.00079 in).
16. $\eta = 18.2 \times 10^{-6}$ N s/m^2.
17. $QF = \dfrac{q\eta}{P_s h_o^3} \cdot \dfrac{a}{L} \cdot \dfrac{L}{D} \cdot \dfrac{1}{K_{go}} \cdot \dfrac{2p_a}{p_r + p_a} = 0.43$ and $QF2 = 0.175$ for $10°$.

18. $q = \dfrac{2 \times 0.5 \times 10^6 \times 20^3 \times 10^{-18} \times 0.5}{0.0000182 \times 0.25 \times 1} \times \dfrac{0.351 + 0.101}{2 \times 0.101} \times 0.43 = 0.000846 \text{ m}^3/\text{s}$

    $q = 0.846$ l/s total free air for two cones.
19. $\lambda_r = [(2 \times 464)/(20 \times 10^{-6})] \times \cos 10° = 45.7$ MN/m (0.26 $\times 10^6$ lbf/in).
20. $\lambda_a = [(3.3 \times 200)/(20 \times 10^{-6})] \times \sin 10° = 5.78$ MN/m (0.033 $\times 10^6$ lbf/in).
21. $H_p = 0.101 \times 10^6 \times 0.000846 = 85.5$ W.
22. $A_f = \pi \times 0.0566 \times 0.0566 = 0.01$ m^2. $U = \pi \times 0.0566 \times 100 = 17.78$ m/s.
    $H_f = (0.000018 \times 0.01 \times 17.78^2)/(20 \times 10^{-6}) = 2.88$ W.
23. $H_t = 85.5 + 2.88 = 88.4$ W.
24. $q_{restrictor} = 0.000846/48 = 0.0000176$ m^3/s free air at large end
    $q_{restrictor} = 2 \times 0.0000176 \times 0.175/0.43 = 0.0000143$ m^3/s at small end.
25. See design procedures for restrictors in Chapter 5.

■

References

Aston, R. L., O'Donoghue, J. P., & Rowe, W. B. (1970). Design of conical hydrostatic journal bearings. *Machinery and Production Engineering, 116,* 250–254, (2988; 18 Feb).

Aston, R. L., O'Donoghue, J. P., & Rowe, W. B. (1971). Hydrostatic bearings for combined radial and axial thrust applications. *Proceedings of the Conference on Externally Pressurized Bearings,* London, Paper C28, pp. 228–244. Institution of Mechanical Engineers.

Rowe, W. B. (1967). Experience with four types of grinding machine spindles. *Proceedings of the 8th International MTDR Conference,* University of Manchester, pp. 453–477. Oxford: Pergamon Press.

CHAPTER 13

Spherical Bearings

Summary of Key Design Formulae

$$W = P_s D^2 \cdot \overline{W} \qquad\qquad \text{Radial load}$$

$$T = P_s D^2 \cdot \overline{T} \qquad\qquad \text{Axial load}$$

$$q = \frac{P_s h_o^3}{\eta} \cdot \beta \overline{B} \qquad\qquad \text{Hydrostatic flow}$$

$$q = \frac{P_s h_o^3}{\eta} \cdot K_{go} \overline{B} \cdot \frac{p_r + p_a}{2 p_a} \qquad\qquad \text{Aerostatic flow}$$

$$\lambda_r = \frac{2 P_s D^2}{h_o} \cdot \overline{W}_{0.5} \qquad\qquad \text{Radial stiffness}$$

$$\lambda_a = \frac{2 P_s D^2}{h_o} \cdot (\overline{T}_{0.5} - \overline{T}_0) \qquad\qquad \text{Axial stiffness}$$

13.1 Application

Spherical bearings tolerate tilt and misalignment and can therefore be employed as pivot bearings or as self-aligning bearings. Hydrostatic or aerostatic spherical bearings may be ideal for measuring instruments where extreme movement accuracy is required. The aerostatic spherical bearing is particularly good for measuring instruments due to extremely low friction. The load capacity of an aerostatic bearing is often lower than for a hydrostatic bearing since lower supply pressure is usually employed. However, load capacity is usually ample for a measuring instrument where the main load is likely to be the weight of the moving parts. It should also be noted that aerostatic flow is increased by the extra pressure terms as given above.

Some examples of hydrostatic spherical bearing configurations are shown in Figure 13.1. The single central recess bearing shown in (a) is useful for reacting substantial constant thrust loads but is poor if the load is not directed towards the central recess. The annular recess bearing in (b) allows a shaft to pass through the bearing and the multi-recess bearing in (c) also supports

Hydrostatic, Aerostatic and Hybrid Bearing Design.
DOI: 10.1016/B978-0-12-396994-1.00013-9

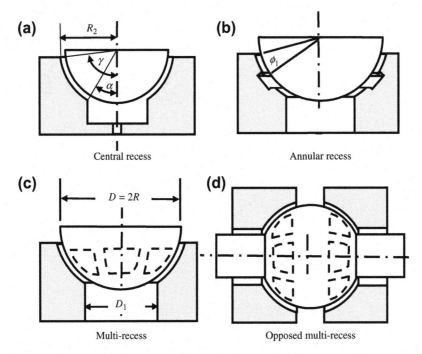

Figure 13.1: Spherical Bearing Recess Patterns.

side loads. The opposed multi-recess configuration in (d) is for combined axial and radial load-ings as required for spindle main bearings.

The configurations in Figure 13.1 are unsuitable for aerostatic spherical bearings without modification to eliminate or substantially reduce recess volumes. Recesses reduce dynamic stiffness of aerostatic bearings and risk pneumatic hammer, as described in early chapters. Chapter 4 gave examples of plane pads modified by employing virtual recesses (see Figures 4.1, 4.7, and 4.13). A "virtual recess" is where a land area is surrounded by a ring of jets or slots to enclose a high-pressure region. Very small and very shallow pockets or recesses can be employed but the total air volume in the bearing must be minimized taking account of the advice in Chapter 5. Figure 4.7 shows a spherical bearing with central admission modified for aerostatic operation by means of a ring of jets or slots to form a virtual recess.

Comparative load capacities for spherical and other types of spindle bearing arrangement are given in Section 1.4: the radial load capacity of a spherical bearing is approximately half that of a Yates bearing and less than one-third that of a cylindrical bearing. Axial thrust load capacity is also low due to the spherical geometry. A face-to-face configuration is shown in Figure 1.9. The problem of face-to-face arrangements is discussed in Section 12.1 in relation to conical bearings. It is preferable to combine a face-to-face opposed spherical main bearing, as shown in Figure 13.1d, with a smaller cylindrical bearing at the other end of the spindle. This arrangement gives better radial load capacity.

Spherical surfaces are more difficult to machine than plane, cylindrical and conical surfaces, and are likely to be more expensive than other configurations. The designer will give careful consideration as to whether requirements justify spherical bearings. Information on a manufacturing technique and further design data are given by Rowe and Stout (1971).

13.2 Basic Parameters

Main design parameters are listed in the "key design formulae" above. Data values are given in charts and tables.

Geometry is defined relative to the principal axis:

D, R	sphere diameter and radius
D_1, R_1	bearing smaller diameter and radius
R_2	bearing larger radius
h_o	concentric film thickness
$\beta = K_{go}$	concentric pressure ratio in a recess P_r/P_s
α and γ	angles that define pad extent as in Figure 13.1
ϕ_l	angle that defines land width for outflow
ϕ_b	angle that defines land width for inter-recess flow
ε	eccentricity ratio e/h_o.

Basic Geometry

Geometry of a recessed spherical bearing is shown in Figure 13.1. The figures show a bearing having either one or six recesses. Each recess is supplied from a constant supply pressure by way of an individual flow restrictor. A separate restrictor is employed for each recess, as for all multi-recess bearings.

Recommended design combinations based on a pressure ratio $\beta = 0.5$ are given in Table 13.1.

Displacement Values and Minimum Film Thickness

Displacement is simply denoted by eccentricity ratio $\varepsilon = e/h_o$ and attitude angle ϕ as in Figure 13.2. Minimum film thickness is therefore $h_{min} = (1 - \varepsilon) \cdot h_o$. When axial loads T and radial loads W are employed together, it is suggested that partial spherical bearings should be designed for eccentricity ratio less than or equal to $\varepsilon = 0.5$.

Table 13.1: Recommended designs, pressure ratio $\beta = 0.5$

Recess Pattern	α	γ	ϕ_1	W/P_sD^2	T/P_sD^2	$\beta\bar{B}$
Central	35°	70°	35°	0	0.23	0.33
Annular	30°	70°	7.5°	0	0.2	2.94
Multi-recess	50°	86°	7.2°	0.09	0.11–0.18	3.6
Opposed bearings	50°	86°	7.2°	0.18	0.06	7.2

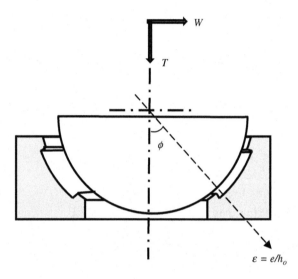

Figure 13.2: Load and Eccentricity Directions.

Pressure Ratio and Restrictor Design

Capillaries feeding each recess should be of such a length and diameter that recess pressure in the concentric condition is 0.4–0.6 times the supply pressure, i.e. $0.4 < \beta < 0.6$.

After designing a bearing, flow will be known for each recess. This value can be used to design the capillaries using the procedure in Appendix 5.A.

Land-Width Ratios

Land width for an annular recess bearing or a multi-recess bearing is shown in Figure 13.1 and defined by the angle ϕ_1. The physical width of the land is $R \cdot \phi_1$. A reasonable value for moderate hydrostatic flow is $\phi_1 = 0.2(\gamma - \alpha)$. For an aerostatic bearing, employing a virtual recess, a reasonable value is $\phi_1 = 0.33(\gamma - \alpha)$.

The inter-recess land width for a six-recess hydrostatic bearing is defined by the inter-recess land-width angle ϕ_b subtended at the center of the sphere. A reasonable range for the inter-recess land is $\phi_b = 10-20°$.

13.3 Central-Recess Spherical Bearing

A single central-recess bearing is usually employed to carry a constant load. At high loads, a central-recess bearing exhibits negative stiffness and may collapse. However, if the load is constant and does not exceed the maximum value, this problem does not arise. It is recommended that this bearing is designed for a pressure ratio $\beta = 0.5$, which means that the bearing will operate near the maximum stiffness region. Recommended design values for spherical bearings are given in Table 13.1. Computed data for a central-recess bearing were given in Section 4.7, together with an example of the basic calculation.

$$\overline{T} = \frac{T}{P_s D^2} = \frac{\beta \pi \sin^2 \gamma}{4} \cdot \overline{A} \qquad \text{Axial thrust load} \qquad (13.1)$$

$$\frac{q\eta}{P_s h_o^3} = \beta \cdot \overline{B} \qquad \text{Hydrostatic flow}$$

$$\frac{q\eta}{P_s h_o^3} = K_{go} \cdot \overline{B} \cdot \frac{p_{ro} + p_a}{2p_a} \qquad \text{Aerostatic flow} \qquad (13.2)$$

where $\beta = K_{go} = P_{ro}/P_s$ is the concentric gauge pressure ratio, and where p_{ro} and p_a are absolute gas pressures.

$$S_h = \frac{\eta N}{P_s} \left(\frac{D}{2h_o} \right)^2 \qquad \text{Speed parameter} \qquad (13.3)$$

The optimum value of S_h is given by $S_{ho} = (1/4\pi)\sqrt{\beta \overline{B}/\overline{A}_f}$, where the friction area factor is

$$\overline{A}_f = \frac{A_f}{D^2} = \frac{\pi}{2} \left[(1 - \cos\gamma) - \frac{3}{4}(1 - \cos\alpha) \right] \qquad (13.4)$$

In equation (13.4), the symbols α and γ denote the bearing shape. There is an exact solution for the flow. Bearing load and flow factors can be read from Figure 4.7 or can be calculated from

$$\overline{A} = \frac{1 - (\sin\alpha/\sin\gamma)^2}{2\log_e(\sin\gamma/\sin\alpha)} \qquad (13.5)$$

$$\overline{B} = \frac{\pi}{6\log_e[(\tan\gamma/2)/(\tan\alpha/2)]} \qquad (13.6)$$

The following is an example illustrating the design of a partial spherical hydrostatic bearing with a central recess. An aerostatic bearing follows a similar procedure, as illustrated by Example 13.2.

■ *Example 13.1 A Central Recess Hydrostatic Spherical Bearing*

1. Specify: axial thrust, $T = 1067$ N (240 lbf).
2. Specify: diameter, $D = 5.08$ cm (2 in).
3. Specify: bearing angles, $\alpha = 35°$, $\gamma = 70°$.
4. Specify: rotational speed, $N = 25$ rev/s.
5. Read: $\bar{A} = 0.67$ from Figure 4.7. $\bar{T} = \dfrac{0.5\pi \sin^2(70°)}{4} \times 0.67 = 0.23$.
6. Calculate: supply pressure from

$$P_s = \frac{T}{D^2 \bar{T}} = \frac{1067}{0.0508^2 \times 0.23} = 1.8 \text{ MN/m}^2 \ (261 \text{ lbf/in}^2)$$

7. Select: radial clearance, $h_o = 33$ μm (0.0013 in).
8. Read: $\bar{B} = 0.67$ from Figure 4.7 and $\beta \cdot \bar{B} = 0.5 \times 0.67 = 0.33$.
9. If $N = 0$, specify: dynamic viscosity η and go to step 14.
10. Calculate: friction area factor:

$$\bar{A}_f = \frac{\pi}{2}[(1 - \cos 70°) - \frac{3}{4}(1 - \cos 35°)] = 0.82$$

11. Calculate: the optimum speed parameter from $S_{ho} = (1/4\pi)\sqrt{\beta \bar{B}/\bar{A}_f}$

$$= (1/4\pi)\sqrt{0.33/0.82} = 0.05.$$

12. Calculate: viscosity from $\eta = (P_s/N)(2h_o/D)^2 \cdot S_h$.
    $\eta = [(1.8 \times 10^6)/25] \ [(2 \times 33 \times 10^{-6})/0.0508]^2 \times 0.05$
    $\quad = 0.0061$ N s/m^2 or 6.1 cP.
    If η is too high, choose a convenient value and go to step 14.
    If η is too low, choose the minimum acceptable value.
13. Calculate: new radial clearance to suit new viscosity from

$$h_o^2 = \frac{\eta N}{P_s}\left(\frac{D}{2}\right)^2 \frac{1}{S_h} \quad \text{(not necessary in this example)}$$

14. Calculate: flow from equation (13.2):

$$q = \frac{1.8 \times 10^6 \times 33^3 \times 10^{-18}}{0.0061} 0.33 = 3.5 \times 10^{-6} \text{ m}^3/\text{s or 0.21l/min}$$

15. Calculate: pumping power from $H_p = P_s q = 1.8 \times 10^6 \times 3.5 \times 10^{-6} = 6.3$ W.
16. Calculate: friction power from $H_f = H_p = 6.3$ W, where $K = H_f/H_p = 1$.
17. Calculate: total power $H_t = H_p + H_f = 6.3 + 6.3 = 12.6$ W.

18. Calculate: maximum temperature rise for a single pass of the lubricant through the bearing from $\Delta T = H_t/c\rho q = 12.6/(2120 \times 855 \times 3.5 \times 10^{-6}) = 2 \,^{\circ}C$, where the lubricant properties are specific heat capacity $c = 2120$ J/kg K and density $\rho = 855$ kg/m^3. The maximum temperature rise can also be simply calculated from $\Delta T = [P_s(1 + K)]/c\rho$, where K equals the ratio friction power to pumping power.

The calculation of the restrictors follows the procedure in Chapter 5.

■

■ *Example 13.2 A Virtual Central-Recess Aerostatic Spherical Bearing*

1. Specify: axial thrust, $T = 250$ N (56.2 lbf).
2. Specify: diameter, $D = 5.08$ cm (2 in).
3. Specify: bearing angles, $\alpha = 35°$, $\gamma = 70°$.
4. Specify: rotational speed, $N = 15$ rev/s.
5. Read: $\overline{A} = 0.67$ from Figure 4.7.

$$\overline{T} = \frac{0.5\pi\sin^2(70°)}{4} \times 0.67 = 0.23$$

6. Calculate: gauge and absolute supply pressures from

$P_s = T/D^2\overline{T} = 250/(0.0508^2 \times 0.23) = 0.421$ MN/m^2 (61.1 lbf/in^2)
$p_s = P_s + p_a = 0.421 + 0.101 = 0.522$ MN/m^2 (75.7 lbf/in^2).

7. Calculate: gauge entry pressure to bearing from $P_{ro} = P_sK_{go} = 0.421 \times 0.5 = 0.2105$ MN/m^2 (30.5 lbf/in^2).
8. Calculate: absolute entry pressure from $p_{ro} = P_{ro} + p_a = 0.2105 + 0.101 = 0.311$ MN/m^2 (45.2 lbf/in^2) for ambient pressure $p_a = 0.101$ MN/m^2.
9. Specify: radial clearance, $h_o = 20$ μm (0.00079 in).
10. Read: $\overline{B} = 0.67$ from Figure 4.7. The flow factor $K_{go}\overline{B} = 0.5 \times 0.67 = 0.33$.
11. Specify: dynamic viscosity η. For air $\eta = 0.0000182$ N s/m^2 at 18 °C.
12. Calculate: free air flow from equation (13.2):

$$q = \frac{0.421 \times 10^6 \times 20^3 \times 10^{-18}}{0.0000182} 0.33 \times \frac{0.311 + 0.101}{2 \times 0.101}$$

$$= 125 \times 10^{-6} \text{ m}^3/\text{s} = 0.125 \text{ l/s or } 7.47 \text{ l/min}$$

13. Calculate: pumping power from

$$H_p = p_aq_a\ln\left(\frac{p_s}{p_a}\right) = 0.101 \times 10^6 \times 125 \times 10^{-6} \times \ln\left(\frac{0.522}{0.101}\right) = 20.7 \text{ W}$$

Calculation of restrictors follows the procedures in Chapter 5.

■

13.4 Annular-Recess Spherical Bearing

An annular-recess spherical bearing is illustrated in Figure 13.1b. Design follows very similar lines to the previous example. Basic bearing geometry is represented by angles α and γ as previously. However, axial flow land width is now represented by the angle ϕ_l and inter-recess land width by the angle ϕ_b. The suggested value of the inter-recess land-width angle is $\phi_b = 20°$. Other recommended values are specified in Example 13.3.

The annular-recess spherical bearing has a relatively short axial length. The suggested number of recesses is $n = 6$, although $n = 4$ or 5 are reasonable geometries. However, a six-recess bearing offers better load support and stiffness.

The bearing can be designed using the same procedure as in the previous example by substituting the following expressions for the load, flow, and friction area factors:

$$\overline{T} = \frac{\beta\pi}{4}\left\{\frac{\sin^{2\gamma}\left[1 - (\sin(\gamma - \phi_l)/\sin\gamma)^2\right]}{2\log_e(\sin\gamma/\sin(\gamma - \phi_l))} - \frac{\sin^2(\alpha + \phi_l)\left[1 - (\sin\alpha/\sin(\alpha + \phi_l))^2\right]}{2\log_e(\sin(\alpha + \phi_l)/\sin\alpha)}\right\}$$

(13.7)

$$\overline{B} = \frac{\pi}{6\log_e\left[\tan\dfrac{\gamma}{2}\Big/\tan\dfrac{\gamma - \phi_l}{2}\right]} + \frac{\pi}{6\log_e\left[\tan\dfrac{\alpha + \phi_l}{2}\Big/\tan\dfrac{\alpha}{2}\right]}$$

(13.8)

$$\overline{A}_f = \frac{A_f}{D^2} = \frac{\pi}{2}\left[\left(\cos\alpha - \cos\gamma\right) - \frac{3}{4}(\cos(\alpha + \phi_l) - \cos(\gamma - \phi_l))\right]$$

(13.9)

The following abbreviated hydrostatic example shows the calculation of load, flow, and friction area factors for the suggested geometry. The annular-recess bearing can be designed for aerostatic operation by replacing the large recess with a virtual recess formed by two rings of jets or slots as in previous examples. The land-width angle can be increased to $\phi_l = 10°$ if desired to modify the following example for aerostatic operation. This will slightly reduce flow. Aerostatic design is similar to Example 13.2 for an aerostatic central-recess bearing except that $\overline{T}$ and $\overline{Q}$ are calculated as in Example 13.3.

■ **Example 13.3 Annular-Recess Bearing for α = 30°, γ = 70°, and φ$_l$ = 7.5°**

Evaluate the load, flow, and friction factors for the specified geometry.

1. For a concentric pressure ratio $\beta = 0.5$,

$$\overline{T} = \frac{0.5\pi}{4} \left\{ \begin{array}{c} \dfrac{\sin^2(70^\circ)\left[1 - (\sin(62.5^\circ)/\sin(70^\circ))^2\right]}{2\log_e((\sin(70^\circ)/\sin(62.5^\circ))} \\[4mm] \dfrac{\sin^2(37.5^\circ)\left[1 - (\sin(30^\circ)/\sin(37.5^\circ))^2\right]}{2\log_e(\sin(37.5^\circ)/\sin(30^\circ))} \end{array} \right\}$$

$$\overline{T} = 0.207$$

2. $\beta\overline{B} = 0.5\pi/\{6\log_e[(\tan(70^\circ/2))/(\tan(62.5^\circ)/2))]\} + 0.5\pi/\{6\log_e[(\tan(37.5^\circ/2))/(\tan(30^\circ/2))]\} = 2.94$.
3. $\overline{A}_f = (\pi/2)[(\cos(30^\circ) - \cos(70^\circ)) - {}^3/_4(\cos(37.5^\circ) - \cos(62.5^\circ))] = 0.43$.

∎

13.5 Single Multi-Recess Spherical Bearings

The geometry is described in Figure 13.1c. The bearing angles are α, γ, ϕ_l, and ϕ_b. Since the purpose of the multi-recess bearing is to withstand radial loadings normal to the principal axis, the recommended bearing angles have been increased to $\alpha = 50^\circ$ and $\gamma = 86^\circ$. The increased angles improve radial loadings that can be safely applied. The recommended concentric pressure ratio is $\beta = 0.5$ and bearing data are given for six-recess bearings.

The geometry shown in Figure 13.1c is unsuitable for aerostatic operation. It is better to employ two rings of slot or jet restrictors as suggested for the aerostatic annular-recess bearing. The advantage of this arrangement is that multiple rings of restrictors provide both radial load support and thrust load support. Load support is similar to the multi-recess spherical bearing.

Flow for single multi-recess spherical bearings is given in Figure 13.3 for eccentricity ratios up to 0.75. Flows are given for $\phi_l = 3.6^\circ$. However, flows can easily be estimated for larger land width since flow is inversely proportional to land width for this geometry. Flow is presented for a concentric pressure ratio $\beta = 0.5$. Flow for other pressure ratios can be obtained from the figure since concentric flow increases in direct proportion with β. For example, for $\beta = 0.25$, concentric flow is half the value from Figure 13.3.

Combined thrust and radial loads for a single multi-recess spherical bearing are given in Figure 13.4.

13.6 Opposed Multi-Recess Spherical Bearings

Multi-recess spherical bearings may be applied in a single-bearing arrangement or in an opposed-bearing configuration. When spherical bearings are employed for the main bearings

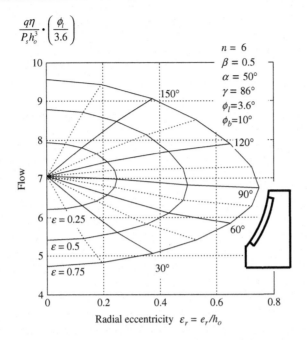

Figure 13.3: Flow and Eccentricity Ratio for a Single Multi-Recess Spherical Bearing.

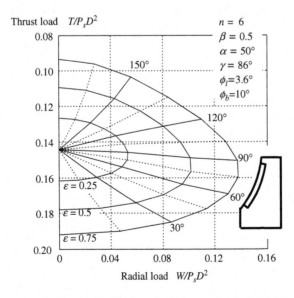

Figure 13.4: Thrust Load *T* and Radial Load *W* for a Single Multi-Recess Spherical Bearing.

on a spindle, they will usually be combined in the opposed-bearing form. When the two spherical bearings are located close together in a true self-aligning arrangement, it is usual to add a cylindrical journal bearing, as illustrated in Figure 13.5.

It is recommended that six recesses are included in each of the two opposed bearings and the concentric pressure ratio should be $\beta = 0.5$. Thrust loads and radial loads are given in Figure 13.6 for a recommended arrangement suitable for longer shafts.

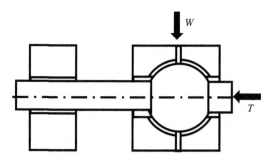

Figure 13.5: Opposed Spherical Pads Combined with a Cylindrical Journal Bearing.

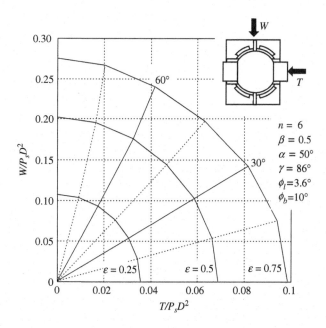

Figure 13.6: Load Parameters for an Opposed Spherical Bearing Arrangement.

■ *Example 13.4 An Opposed Multi-Recess Spherical Bearing:* $\alpha = 50°$, $\gamma = 86°$, $\phi_l = 7.2°$

This procedure may also be used as a guide for single spherical bearings.

1. Radial load, $W = 1800$ N (405 lbf).
2. Axial thrust load, $T = 1200$ N (270 lbf).
3. Shaft diameter of journal, $D_j = 70$ mm (2.76 in).
4. Rotational speed, $N = 20$ rev/s.
5. Sphere geometry, $\alpha = 50°$, $\gamma = 86°$, $\phi_l = 7.2°$ (see Table 13.1 and Figure 13.1).
6. Sphere diameter $D = D_j/\sin \alpha = 70/0.766 = 91.38$ mm (3.6 in).
7. Concentric pressure ratio $\beta = 0.5$.
8. Minimum supply pressure to support radial load from Table 13.1,
   $P_s = 1800/(0.09138^2 \times 0.18) = 1.198$ MN/m^2 (174 lbf/in^2).
9. Minimum supply pressure to support the thrust load from Table 13.1,
   $P_s = 1200/(0.09138^2 \times 0.06) = 2.395$ MN/m^2 (347 lbf/in^2).
10. Supply pressure, $P_s = 2.4$ MN/m^2.
11. Radial clearance (initial selection), $h_o = 33$ μm (0.0013 in).
12. Read concentric flow factor from Table 13.1 for the bearing pair, $\beta \bar{B} = 7.2$.
13. If $N = 0$, specify absolute viscosity and go to step 19.
14. Calculate effective friction area, A_f, for bearing pair from $A_f = A - 3A_r/4$:

$$A = \pi D^2 (\cos \alpha - \cos \gamma) = \pi \times 0.09138^2 \times (0.6428 - 0.0698) = 0.015 \text{ m}^2$$

$$A_r = \pi D^2 (\cos 57.2 - \cos 78.8)\left(1 - \frac{n\phi_b}{360}\right)$$

$$= \pi \times 0.09138^2 (0.5417 - 0.1942)\left(1 - \frac{6 \times 10}{360}\right) = 0.01 \text{ m}^2$$

$$A_f = 0.015 - 0.75 \times 0.01 = 0.0075 \text{ m}^2 \ (11.62 \text{ in}^2)$$

15. Calculate average sliding speed at the average angle $(\alpha + \gamma)/2 = (50 + 86)/2 = 68°$. $U = \pi D N \sin \theta = \pi \times 0.09138 \times 20 \times \sin 68° = 5.32$ m/s (210 in/s).
16. Calculate optimum viscosity from

$$\eta = \frac{P_s h_o^2}{U}\sqrt{\frac{\beta \bar{B}}{A_f}} = \frac{2.4 \times 10^6 \times 33^2 \times 10^{-12}}{5.32}\sqrt{\frac{7.2}{0.0075}}$$

$$= 0.015 \text{ N s/m}^2 \ (2.18 \times 10^{-6} \text{ reyns})$$

17. If η is too high, choose a convenient value and go to step 19. If η is too low, choose the minimum acceptable value

18. Calculate a new value of radial clearance from $h_o^2 = (\eta U / P_s) \sqrt{A_f / \beta \overline{B}}$.

19. Calculate concentric flow rate from

$$q = \frac{P_s h_o^3}{\eta} \beta \overline{B} = \frac{2.4 \times 10^6 \times 33^3 \times 10^{-18}}{0.015} \times 7.2 = 41.4 \times 10^{-6} \text{ m}^3/\text{s} \ (2.48 \text{ l/min})$$

20. Calculate pumping power from $H_p = P_s q = 2.4 \times 10^6 \times 41.4 \times 10^{-6} = 99.36$ W.

21. Calculate friction power from $H_f = (\eta A_f U^2)/h_o$
    $= (0.015 \times 0.0075 \times 5.32^2)/(33 \times 10^{-6}) = 96.5$ W.

22. Total power $H_t = H_p + H_f = 196$ W (0.26 hp).

23. Maximum temperature rise for 1 pass of the lubricant where $c = 2120$ J/kg K and $\rho = 855$ kg/m^3 is given by $\Delta T = H_t / c \rho q = 196/(2120 \times 855 \times 41.4 \times 10^{-6}) = 2.61$ K.

24. Concentric flow rate through each restrictor $q_r = q/n = 41.4 \times 10^{-6}/12 = 3.45 \times 10^{-6}$ m^3/s. This value should be used to design the restrictors.

∎

Reference

Rowe, W. B., & Stout, K. J. (1971). Design data and a manufacturing technique for spherical hydrostatic bearings in machine tool applications. *International Journal of Machine Tool Design and Research, 11*, (4; Oct.), 293–307.

CHAPTER 14

Dynamics

Summary of Key Design Formulae

$$\lambda_{hs} = \frac{P_s LD}{h_o} \cdot \overline{\lambda}_{hs} \qquad \text{Hydrostatic stiffness}$$

$$\lambda_{ae} = \frac{P_s LD}{h_o} \cdot \overline{\lambda}_{ae} \qquad \text{Aerostatic stiffness}$$

$$\lambda_{hd} = \frac{P_s LD}{h_o} \cdot \overline{\lambda}_{hd} \qquad \text{Hydrodynamic stiffness}$$

$$C_{sq} = \frac{P_s LD}{h_o} \cdot \overline{C}_{sq} \qquad \text{Squeeze film damping}$$

14.1 Introduction

The bearing film is only one element in a total machine system. The dynamic behavior of a system involves the forces and vibrations of all the machine elements. Prediction of system behavior usually involves describing the load/deflection characteristics of each element and building up a composite picture of the combination. Experimental determination of system behavior is the inverse of the synthetic approach used for prediction. In experiments, it is necessary to study the resulting load/deflection characteristics of the overall system, in which it is not usually a simple matter to distinguish the effect of one element such as the bearing film. The designer has no choice since the machine has yet not been built. The designer is therefore forced to take the synthetic approach. The designer starts with limited knowledge of how a system behaves and sets out to design a machine for one or more of the following objectives:

1. To avoid critical or resonant frequencies coinciding with drive and spindle frequencies.
2. To achieve high rigidity, i.e. applied forces cause small deflections.
3. To achieve high damping, i.e. vibration work energy should be dissipated as heat energy and not transmitted as mechanical vibration through the system.
4. To avoid the onset of whirl instability at high rotational speeds in journal bearing design.

Hydrostatic, Aerostatic and Hybrid Bearing Design.
DOI: 10.1016/B978-0-12-396994-1.00014-0

In order to make predictions, a designer often approximates a theoretical model of a proposed system. The elements of a system are modeled in terms of spring stiffness, damping, and mass distribution.

14.2 Static Loading

In Figure 14.1, a large work-piece of weight W will cause a machine table to bend and the bearing film thickness to decrease. Quite often the deflection of the machine structure is of the same order as the deflection of the bearing lubricant film. This is usually considered to be a feature of good design. A machine is a chain of elements in which no element should be a weak link and in which no element should be over-designed. The deflections of the elements in this case add up according to the rule for springs in series:

$$x = \frac{W}{\lambda_k} + \frac{W}{\lambda_b} \tag{14.1}$$

where λ_k is the structural stiffness and λ_b is the bearing film stiffness.

If the table in Figure 14.1 is supported on four pads and is loaded centrally, the load will be divided between the four pads, while the deflection x_b in each pad will be the same according to the rule for springs in parallel:

$$x = \frac{W}{\lambda_k} + \frac{W}{\lambda_b} \tag{14.2}$$

In this case, the resulting bearing stiffness of the bearing pads acting as for springs in parallel is the sum of the pad stiffnesses:

$$\lambda_b = 4x\frac{\lambda_b}{4} \tag{14.3}$$

Stiffness is correctly defined as the slope of the load–deflection curve. In bearing films, bearing film stiffness is not constant with increasing load and therefore load divided by deflection is at best an indication of an average stiffness value.

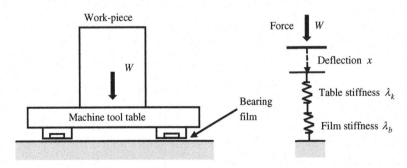

Figure 14.1: Static Loading Representation of a Structure Including Bearings.

14.3 Dynamic Loading

When an applied force F varies with time, the system is dynamic. At any instant an applied force on a system will be reacted by a spring force, a damping force, and inertia due to the mass. For the mass–spring–dashpot system shown in Figure 14.2, an equation of motion describes the force balance:

$$m\ddot{x} + C\dot{x} + \lambda x = F \tag{14.4}$$

In equation (14.4), F is a function of time expressed by the statement $F = F(t)$ and x is a function of time so that $x = x(t)$. For convenience, these statements are implicit in dynamic equations and we refer to F and x in the same way as for static parameters. This simplifies the presentation of dynamic relationships. Concise notation for the first time-derivative of deflection is the velocity $\dot{x} = dx/dt$ and the second time-derivative is the acceleration $\ddot{x} = d^2x/dt^2$.

High mass and low stiffness associated with low damping lead to resonance at a low characteristic frequency. Sufficient damping will reduce or even eliminate resonance.

The following sections introduce the origins and prediction of the stiffness and damping characteristics of hydrostatic, hybrid, and aerostatic bearing films.

Before going on, a note of caution should be sounded. Rapidly increasing loads followed by rapidly reducing loads may lead to three effects that invalidate calculations. First, a liquid film may cavitate if dynamic loading is sustained at large amplitudes and hence produce bubbles, thereby causing the liquid film to become much more compressible than normal. Second, heating may occur, leading to variable viscosity effects. This effect will be more pronounced with high-viscosity liquids than with low-viscosity liquids. A third effect is the nonlinear load versus deflection characteristic of hydrostatic films. For simple-linear calculations, vibration forcing amplitudes must be small in comparison with the static loads.

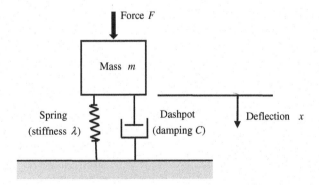

Figure 14.2: Spring–Mass–Dashpot System.

For gas film bearings, compressibility is taken into account by incorporating the gas laws into the Reynolds equations. Gas film bearings are much more susceptible to resonance problems than liquid film bearings.

14.4 Squeeze Film Damping

The presence of a liquid film in a machine is a powerful source of vibration damping, where the vibrations act to squeeze the liquid through a narrow gap. The squeeze effect in a hydrostatic bearing is a well-known mechanism that operates in the same way as a hydraulic damper (Figure 14.3a, b). In both cases, a volume of liquid has to be displaced at a flow rate equal to the projected area of the moving surface multiplied by the squeeze velocity, V, i.e. $q_{squeeze} = A \cdot V$.

The displaced liquid is squeezed through thin channels that resist the flow and create back-pressure in the entry recess. In the case of a hydrostatic pad, some of the squeeze flow adds to the normal leakage flow leaving the pad. The remainder tries to flow back through the control restrictor, thus subtracting from the normal flow into the recess.

The effects of squeeze velocity on pressures in a hydrostatic pad are shown in Figure 14.3c. Recess pressure is increased, which has the effect of increasing pressure on the lands. In

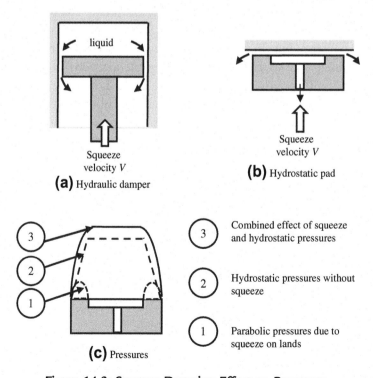

(a) Hydraulic damper

(b) Hydrostatic pad

(c) Pressures

3 — Combined effect of squeeze and hydrostatic pressures

2 — Hydrostatic pressures without squeeze

1 — Parabolic pressures due to squeeze on lands

Figure 14.3: Squeeze Damping Effect on Pressures.

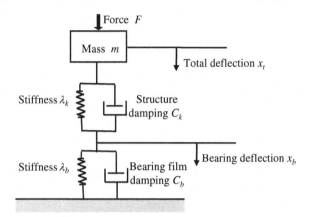

Figure 14.4: Approximate Low-Frequency Model of a Hydrostatic Bearing and a Simple Structure.

addition, squeeze pressures are generated on the lands that, in the absence of a recess pressure, would give a parabolic distribution. The resultant pressure distribution is the combination of these two sets of pressures.

Squeeze damping is the main source of damping in a machine containing a hydrostatic bearing as long as the bearing stiffness is not substantially higher than the structural stiffness. This may be deduced as follows, with reference to Figure 14.4. A force F applied to the mass will cause it to deflect to a new position. The total movement x_t is the sum of the deflections in the structure of the machine x_k and the deflection across the lubricant film x_b:

$$x_t = x_k + x_b \tag{14.5}$$

If the bearing stiffness is much greater than the structural stiffness, most of the deflection occurs in the structure rather than across the lubricant film and deflection of the structure is mainly damped by the structural damping C_k. Very little deflection occurs across the bearing film, which acts like a rigid member. Thus, the consequence of high bearing stiffness is that the high value of squeeze damping in the bearing film becomes ineffective to damp out structural resonances and forced vibrations. Failure to recognize the importance of this point has led some designers to the discovery that their machines behave as lightly damped systems when they were intended to be heavily damped.

14.5 Compressibility in Hydrostatic and Aerostatic Bearings

At low frequencies, compressibility of a liquid is often considered negligible. At higher frequencies, compressibility has the effect of reducing dynamic stiffness and damping of a bearing. Compressibility of a lubricant is defined by the bulk modulus, equation (2.4). Compressibility of a gas is defined by the gas laws.

The relatively large volumes of liquid in a hydrostatic bearing recess and between the recess and the restrictor give rise to a significant compressibility effect on recess pressure. The compressibility of liquid in the thin-film region between the bearing lands is normally of much lower importance due to its relatively small volume.

Compressibility assumes much greater importance in an aerostatic bearing, as mentioned in Section 14.3. In gas bearing design, large recesses are avoided since large recesses lead to reduced dynamic stiffness and risk bearing instability due to pneumatic hammer. This can be seen from the following simple dynamic model of a thrust pad.

14.6 Dynamic Model of a Thrust Pad

A simple model of a hydrostatic or aerostatic thrust bearing contained in a structure is shown in Figure 14.4. Given values of bearing and structural characteristics such as λ_b, C_b, λ_k, C_k, a designer may use any one of the standard techniques to establish resonant frequency and frequency responses. However, it is not intended to detail these techniques, which are readily available in textbooks on vibrations. The purpose of this section is to introduce the behavioral properties of liquid films in hydrostatic bearings. The first problem is to obtain the characteristics of the liquid film denoted by the values λ_b and C_b. It is also necessary to establish the parameters governing the magnitude of deflection. Subsequently the designer will need to decide whether the magnitudes of λ_b and C_b are appropriate to the requirements of the whole system.

Although Figure 14.4 is an attractive representation of small-amplitude low-frequency vibrations, it is rather oversimplified for an accurate representation of many situations. Furthermore, the designer may wish to incorporate the characteristics of the fluid film into a more complex system. For general usage it is more helpful to examine the fluid-film transfer function. The transfer function defined with reference to Figure 14.5a is

$$TF(\text{bearing film}) = \Theta = \frac{x}{W} \tag{14.6}$$

where x represents the dynamically changing bearing film thickness and W is the dynamically changing bearing film thrust.

The use of transfer functions to analyze responses may be demonstrated for the case of a bearing film model in Figure 14.4. To illustrate the transfer function approach, it will be assumed that the stiffness and damping values are known and are constant with amplitude and frequency. The input thrust W is the bearing thrust due to the film pressures acting on the bearing surface and the output x is the change in bearing film thickness. The equation of motion is

$$W = C_b \frac{\mathrm{d}x}{\mathrm{d}t} + \lambda_b x \tag{14.7}$$

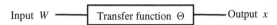

(a) Transfer function representation of film characteristics

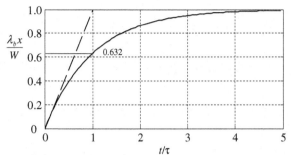

(b) Step response for the TF: $x/W = 1/\lambda_b(1 + \tau D)$

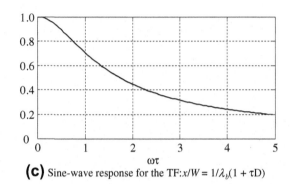

(c) Sine-wave response for the TF: $x/W = 1/\lambda_b(1 + \tau D)$

Figure 14.5: Responses of a Lubricant Film Having a TF: $x/W = 1/\lambda_b(1 + \tau D)$.

The transfer function can be represented by

$$\Theta = \frac{x}{W} = \frac{1}{\lambda_b + C_b D} \tag{14.8}$$

The D operator denotes the operation of differentiation d/d*t*. The transfer function is sometimes given in the form:

$$\Theta = \frac{x}{W} = \frac{1}{\lambda_b}\left(\frac{1}{1 + \tau D}\right) \tag{14.9}$$

where $\tau = C_b/\lambda_b$ is known as the time constant.

The solution of equation (14.9) depends on the initial state of the system and the nature of the input. If the disturbance is a step function and the initial state is steady, the response is as shown in Figure 14.5b.

For sinusoidal forcing from a steady initial state, $W(t) = W\sin \omega t$, the solution is

$$\frac{\lambda_b x}{W} = \frac{\omega \tau}{1 + \omega^2 \tau^2} \cdot e^{-t/\tau} + \frac{1}{\sqrt{1 + \omega^2 \tau^2}} \cdot \sin (\omega t - \phi) \tag{14.10}$$

The steady-state amplitude of the response is given by letting $t = \infty$ when

$$\left|\frac{\lambda_b x}{W}\right| = \frac{1}{\sqrt{1 + \omega^2 \tau^2}} \tag{14.11}$$

The amplitude decreases with frequency, as shown in Figure 14.5c. This is a typical shape of the frequency response characteristic for a hydrostatic bearing excited at low frequencies. A hydrostatic bearing film is well damped due to the squeeze-film effect. The steady-state frequency response could alternatively have been obtained by substituting $j\omega$ for D in equation (14.9). This is a useful technique for extracting the values of λ_b and C_b from the transfer function if required:

$$\frac{1}{\Theta(\omega)} = \frac{W}{x} = \lambda_b + j\omega C_b \tag{14.12}$$

The values of λ_b and C_b can therefore be established from the real (Re) and imaginary (Im) parts of W/x. Using a phase-meter to analyze the response in-phase with the forcing signal and in quadrature with the forcing signal gives the real and imaginary parts at a particular frequency ω:

$$\begin{aligned} \lambda_b &= \frac{W}{x} \qquad \text{(Re)} \\[2mm] C_b &= \frac{1}{\omega}\frac{W}{x} \qquad \text{(Im)} \end{aligned} \tag{14.13}$$

It will be found, in practice, that the values of λ_b and C_b are not exactly constant with frequency and will be found to reduce at higher frequencies. One reason for this is compressibility of the lubricant.

The phase angle ϕ in equation (14.10) is given by

$$\tan\phi = \omega\tau = \omega\frac{C_b}{\lambda_b} \tag{14.14}$$

The following section deals with the prediction of C_b and λ_b for a flat thrust pad with thin lands.

The compressibility of an aerostatic bearing dominates dynamic performance and it is possible to assume for the purposes of predicting a resonant frequency that $C_b = 0$. Damping of an aerostatic bearing is improved if the recess volume of gas is completely eliminated.

Damping arises because gas is forced to flow through the thin bearing film where viscous forces predominate. This is the explanation for the superior dynamic stiffness of slot-restricted bearings without recesses compared with orifice-fed bearings with recesses.

An aerostatic bearing will usually exhibit a resonance at a frequency $\omega_n = \sqrt{\lambda/M}$, where M is the supported mass and λ is the associated static stiffness due to the aerostatic bearing film.

■ *Example 14.1 Resonant Frequency of an Aerostatic Bearing*

In Example 2.2, an aerostatic bearing for a simple circular pad supports a weight of 30.4 N. Calculate the approximate resonant frequency.

The weight of 30.4 N corresponds to a mass of $30.4/9.81 = 3.1$ kg.

1. Specified: mass, $M = 3.1$ kg.
2. Specified: film thickness, $h_o = 10$ μm.
3. Specified: gauge pressure ratio, $K_{go} = 0.5$.
4. Specified: gauge supply pressure from Example 2.2, $P_s = 0.4$ MN/m^2.
5. Specified: absolute entry pressure from Example 2.2, $p_r = 0.3$ MN/m^2.
6. Specified: bearing area from Example 2.2, $A = 0.000707$ m^2.
7. Specified: area factor from Example 2.2, $\overline{A} = 0.215$.
8. Calculate: effective area from $A_e = A \cdot \overline{A} = 0.000707 \times 0.215 = 0.000152$ m^2.
9. Read: aerostatic stiffness factor for capillary control from Figure 7.2, $\overline{\lambda}_{ae} = 0.62$.
10. Calculate: stiffness from

$$\lambda = \frac{P_s A_e}{h} \cdot \overline{\lambda}_{ae} = \frac{0.4 \times 10^6 \times 0.000152}{10 \times 10^{-6}} = 6.08 \text{ MN/m}$$

11. Calculate: resonant frequency from

$$\omega_n = \sqrt{\frac{\lambda}{M}} = \sqrt{\frac{6.08 \times 10^6}{3.1}} = 1400 \text{ rad/s}$$

12. Calculate: resonant frequency in hertz from $f_n = \omega_n/2\pi = 1400/2\pi = 223$ Hz.

■

14.7 Hydrostatic Thrust Pad with Thin Lands

The transfer function for a hydrostatic thrust pad with thin lands may be obtained analytically by considering the dynamic changes of flow into the recess and out from the recess due to changes in recess pressure and the bearing film thickness. Since recess pressure is directly proportional to bearing thrust, this analysis leads straightforwardly to the effects on bearing thrust.

For a capillary-controlled pad:

$$q_{\text{squeeze}} = A_e \frac{dh}{dt}$$

where A_e is the effective area of the pad:

$$q_{\text{compressibility}} = \frac{V}{K} \frac{dp_r}{dt}$$

where V is the volume of liquid enclosed in the bearing and K is the bulk modulus of the liquid:

$$q_{in} = \frac{P_s - p_r}{Z_1} = (P_s - p_r)/(K_c\eta)$$

$$q_{out} = p_r/Z_2 = p_r\frac{\overline{B}h^3}{\eta}$$

Flow into a recess from a restrictor must be equal to flow out of the recess, so that for static loading:

$$\frac{P_s - p_r}{Z_1} = \frac{p_r}{Z_2} \tag{14.15}$$

For dynamic loading, flow equilibrium may be expressed in terms of small perturbations:

$$\frac{dq_{in}}{dp_r}\cdot p = \frac{\partial q_{out}}{\partial p_r}\cdot p + \frac{\partial q_{out}}{\partial h}\cdot h + A_e\frac{dh}{dt} + \frac{V}{K}\frac{dp}{dt} \tag{14.16}$$

where p and h are perturbations of recess pressure and gap. Performing the operations indicated by equation (14.16) leads to

$$-\frac{1}{K_c\eta}p = \frac{\overline{B}h_o^3}{\eta}p + 3p_r\frac{\overline{B}h_o^2}{\eta}h + A_eDh + \frac{V}{K}Dp \tag{14.17}$$

where D stands for the time derivative. The perturbation in recess pressure causes a small variation in the bearing thrust from W_o to $W_o + w$, where $w = pA_e$. Equation (14.17) may therefore be written in terms of w and x, where $x = -h$, to yield a transfer function:

$$\frac{x}{w} = \frac{\dfrac{\overline{B}h_o^3}{\eta} + \dfrac{1}{K_c\eta} + \dfrac{V}{K}D}{A_e\left(3p_r\dfrac{\overline{B}h_o^2}{\eta} + A_eD\right)} \tag{14.18}$$

This equation may be solved for the steady-state frequency response by substituting $D = j\omega$. The real and imaginary parts of w/x are

$$\lambda_b = \frac{\dfrac{3W_o}{h_o}\dfrac{1}{Z_2}\left(\dfrac{1}{Z_1} + \dfrac{1}{Z_2}\right) + \omega^2 A_e^2\dfrac{V}{K}}{\left(\dfrac{1}{Z_1} + \dfrac{1}{Z_2}\right)^2 + \dfrac{\omega^2 V^2}{K^2}} \tag{14.19}$$

$$C_b = \frac{A_e^2\left(\dfrac{1}{Z_1} + \dfrac{1}{Z_2}\right) - 3\dfrac{W_o}{h_o}\dfrac{1}{Z_2}\dfrac{V}{K}}{\left(\dfrac{1}{Z_1} + \dfrac{1}{Z_2}\right)^2 + \dfrac{\omega^2 V^2}{K^2}} \tag{14.20}$$

Stiffness and damping for any value of pressure ratio, $\beta = p_r/P_s$, are given by

$$\lambda_b = \frac{\dfrac{3W_o}{h_o}\left(\dfrac{q}{P_s}\right)^2 \dfrac{1}{\beta^2(1-\beta)} + \omega^2 A_e^2 \dfrac{V}{K}}{\left(\dfrac{q}{P_s}\right)^2 \dfrac{1}{\beta^2(1-\beta)^2} + \dfrac{\omega^2 V^2}{K^2}} \tag{14.21}$$

$$C_b = \frac{A_e^2 \dfrac{q}{P_s} \dfrac{1}{\beta(1-\beta)} - 3\dfrac{W_o}{h_o}\dfrac{V}{K}\dfrac{q}{P_s}\dfrac{1}{\beta}}{\left(\dfrac{q}{P_s}\right)^2 \dfrac{1}{\beta^2(1-\beta)^2} + \dfrac{\omega^2 V^2}{K^2}} \tag{14.22}$$

If the lubricant is incompressible or V is negligible, these equations reduce to

$$\lambda_b = 3\frac{W_o}{h_o}(1-\beta) \tag{14.23}$$

$$C_b = A_e^2\frac{P_s}{q}\beta(1-\beta) \tag{14.24}$$

The responses are given by Figure 14.5, where $\tau = C_b/\lambda_b$, since the stiffness and damping from equations (14.23) and (14.24) are constant and independent of frequency. The stiffness and damping terms derived above describe the dynamic bearing forces that result from the changes in recess pressure. These values are affected by the restrictor resistance as well as the resistance of the bearing lands, assuming the shape of the hydrostatic pressure distribution remains unchanged, as in Figure 14.3c, line 2. Although the squeeze effect has been included in the derivation, parabolic squeeze pressures on thin lands are assumed to have a negligible effect on the pressure distribution shape.

Strictly, even for thin-land bearings, the pressure distribution shape is modified by the parabolic pressures on the lands that are superimposed on the normal pressure distribution, as shown in Figure 14.3c, line 3. The additional damping for a thin land of thickness a and land area A_L is

$$C_2 = \frac{\eta A_L a^2}{h^3} = \text{thin-land damping} \tag{14.25}$$

This value, which may also be employed for long rectangular pads, should be added to the value given by equations (14.20) and (14.21). We may therefore write

$$C_b = C_1 + C_2 \tag{14.26}$$

where C_1 is given by equation (14.22), which includes the effect of compressibility, or by equation (14.24), neglecting compressibility. C_2, given by equation (14.25), is equally applicable for thin-land bearings and for long rectangular bearings. For a circular recessed bearing, C_2 is given by

$$C_2 = \frac{3\pi\eta R_2^4}{2h^3}\left(1 - \left(\frac{R_1}{R_2}\right)^4 - \frac{\left[1 - (R_1/R_2)^2\right]^2}{\log_e(R_2/R_1)}\right) \tag{14.27}$$

When $R_1 = 0.9R_2$, equation (14.25) gives the same result as equation (14.27), i.e. $C_2 = 0.01\eta A_L R_2^2/h^3$. Equation (14.25) therefore provides a useful cross-check. Even with $R_1 = 0.5R_2$, the accuracy of equation (14.25) is within 3%.

■ **Example 14.2 Stiffness and Damping of a Circular Recessed Pad at 20 Hz**

The specified values for a hydrostatic pad are

$R_1 = 18.15$ mm (0.7146 in)
$R_2 = 36.3$ mm (1.43 in)
$A_e = 2.25 \times 10^{-3}$ m^2 (3.49 in^2)
$P_s = 4$ MN/m^2 (580 lbf/in^2)
$p_r = 2$ MN/m^2 (290 lbf/in^2)
$W_o = 4.5$ kN (1012 lbf)
$h_o = 50$ μm (0.002 in)
$\eta = 35$ cP (5.08×10^{-6} reyn)
$q = 5.36 \times 10^{-6}$ m^3/s (0.327 in^3/s)
$K = 2.275$ GN/m^2 (330000 lbf/in^2)
$V = 1.7 \times 10^4$ mm^3 (1.04 in^3).

Solution

$$\omega = 2\pi 20 = 125.6 \text{ rad/s}$$

$$\frac{q}{P_s} = \frac{5.36 \times 10^{-6}}{4 \times 10^6} = 1.34 \times 10^{-12} \text{ m}^5/\text{N s } (5.64 \times 10^{-4} \text{ in}^5/\text{lbf s})$$

$$\beta = p_r/P_s = 2/4 = 0.5$$

From equation (14.21):

$$\lambda_b = \frac{38.79 \times 10^{-16} + 5.968 \times 10^{-16}}{28.73 \times 10^{-24} + 0.8809 \times 10^{-24}} = 0.152 \text{ GN/m } (0.863 \times 10^6 \text{ lbf/in})$$

From equation (14.22):

$$C_1 = \frac{0.2713 \times 10^{-16} - 0.0541 \times 10^{-16}}{28.73 \times 10^{-24} + 0.8809 \times 10^{-24}} = 0.7335 \text{ MN s/m } (4188 \text{ lbf s/in})$$

From equation (14.27):

$$C_2 = \frac{0.035 \times \pi \times (0.0363^2 - 0.01815^2) \times (0.0363 - 0.01815)^2}{50^3 \times 10^{-18}} = 0.2866 \times 10^6 \, \text{MN s/m}$$

$$C_b = 0.7335 \times 10^6 + 0.2866 \times 10^6 = 1.02 \, \text{MN s/m} \, (5824 \, \text{lbf s/in})$$

∎

14.8 Journal Bearings: Equations of Motion

The equations of motion of a complete system will depend on the arrangement of masses, structural linkages, and bearing films. The problem may sometimes be made more manageable by following the procedures outlined in Section 14.6 for a system illustrated in Figure 14.6. In the following discussion, the equations of motion will be considered for the forces acting purely within the bearing film.

The bearing film performance for small perturbations from any value of eccentricity ratio may be described by eight bearing coefficients. To illustrate the eight bearing coefficients, consider a rigid shaft, as in Figures 14.6 and 14.7, supported purely by the oil film that reacts against a rigid bearing housing. Translational movements only are considered (no misalignment tendencies). The steady bearing film force W may be resolved in the x and y directions as W_x and W_y, giving rise to equilibrium values of e and ϕ as shown in Figure 14.7. The system is disturbed by small forces w_x and w_y acting in addition to W_x and W_y. The equations of motion across the bearing film describe the relationships between the small displacements x and y from the equilibrium position and the small disturbing forces giving rise to the displacements.

In xy coordinates (nonrotating):

$$b_{11}\dot{x} + b_{12}\dot{y} + a_{11}x + a_{12}y = w_x \tag{14.28}$$

$$b_{21}\dot{x} + b_{22}\dot{y} + a_{21}x + a_{22}y = w_y \tag{14.29}$$

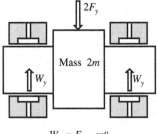

$$W_y = F_y - m\ddot{y}$$

Film force = applied force – acceleration force

Figure 14.6: How the Bearing Film Force May Be Separated From the External Applied Force.

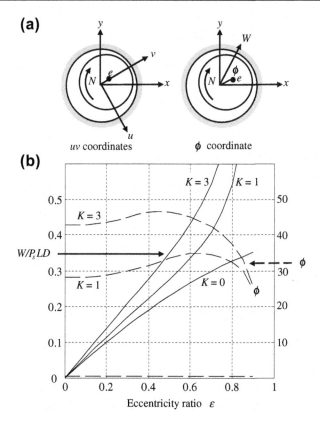

(a)

uv coordinates ϕ coordinate

(b)

Figure 14.7: Bearing Film Coordinates (a) and Static Load Characteristic (b) for $n = 4$, $L/D = 1$, $\beta = 0.5$.

where a_{ij} are the four bearing film stiffness coefficients and b_{ij} are the four bearing film damping coefficients.

The interpretation of the bearing film coefficients is as follows. If a small displacement x from the equilibrium is maintained by the addition of a small steadily applied force w_x in the x direction, then $w_x = a_{11}x$. In general, however, it is not possible to produce a displacement x without also applying a small steadily applied force w_y in the y direction. The required term is $w_y = a_{21}x$. The resultant of the two forces w_x and w_y is

$$w_r = \sqrt{w_x^2 + w_y^2}$$

The direction in which a force w_r must be steadily applied to produce a displacement x is at an angle arctangent (w_y/w_x). By similar reasoning, $a_{12}y$ and $a_{22}y$ are the two forces w_x and w_y that must be applied to produce a displacement y.

The damping coefficients b_{ij} multiplied in a similar way by the instantaneous velocities $\dot{x}$ and $\dot{y}$ yield the instantaneous damping forces.

The stiffness coefficients in *uv* coordinates can be estimated by neglecting compressibility from the static loading characteristics in Figure 14.7:

$$\begin{bmatrix} W_u \\ W_v \end{bmatrix} = \begin{bmatrix} \dfrac{W_v}{e} & \dfrac{\partial W_u}{\partial e} \\ \dfrac{-W_u}{e} & \dfrac{\partial W_v}{\partial e} \end{bmatrix} \begin{bmatrix} u \\ v \end{bmatrix}$$

(14.30)

where $W_u = -W\sin\phi$ and $W_v = W\cos\phi$.

14.9 Journal Bearings in the Concentric Condition

The dynamic analysis of hydrostatic and hybrid bearings is limited, in general, by insufficient reliable data. By limiting consideration to the concentric condition, $e = 0$, it is possible to achieve solutions (Rowe et al., 1979; Rowe, 1980).

Analysis of the concentric condition gives a guide to the critical values required to establish resonant frequencies, the general shape of the displacement/frequency characteristic, and a guide to the possibility of half-speed whirl instability.

Neglecting compressibility, it may be shown that the bearing coefficients for this special case of concentric conditions simplify as follows:

$$a_{11} = a_{22} = \lambda_{hs} \qquad \text{Hydrostatic stiffness} \tag{14.31}$$

$$a_{12} = -a_{21} = -\lambda_{hd} \qquad \text{Hydrodynamic stiffness} \tag{14.32}$$

$$b_{11} = b_{22} = C_{sq} \qquad \text{Squeeze damping} \tag{14.33}$$

$$b_{21} = b_{12} = 0 \tag{14.34}$$

For a small displacement e (see Figure 14.7), the hydrostatic force is independent of speed and acts along the line of eccentricity in the v direction:

$$w_v = \lambda_{hs}e \tag{14.35}$$

The hydrodynamic force for zero eccentricity increases with speed and eccentricity. The hydrodynamic force acts at 90° in the $-u$ direction:

$$w_u = -\lambda_{hd}e \tag{14.36}$$

The resultant of these two forces is

$$w = \sqrt{w_u^2 + w_v^2} \tag{14.37}$$

The attitude angle ϕ between w and e is given by

$$\tan\phi = \frac{-w_u}{w_v} \tag{14.38}$$

For a small velocity $\dot{e}$ in the v direction, there is a squeeze force acting in the same direction:

$$w_v = C_{sq}\dot{e} \tag{14.39}$$

A further simplification arises for the concentric condition as a consequence of symmetry, since Rowe (1980) showed that for this condition squeeze damping is simply related to hydrodynamic stiffness according to

$$C_{sq} = \frac{\lambda_{hd}}{\pi N} \tag{14.40}$$

Hence squeeze damping for this case may be deduced from a steady-loading test for hydrodynamic stiffness.

Analytical expressions λ_{hs}, λ_{hd}, and C_{sq} are given in Tables 14.1 and 14.2 for concentric thin-land recessed bearings and for plain line-entry bearings having either thin or thick lands.

The expressions for plain line-entry bearings are made possible by analyzing both the effects of linear pressure distributions and the effects of parabolic pressure distributions. For hydrostatic stiffness the parabolic pressure distributions detract from stiffness, but for hydrodynamic stiffness the parabolic pressure distributions greatly enhance bearing stiffness and explain the improved performance of plain bearings.

Values of stiffness and damping factors for typical concentric bearings are plotted in Figures 14.8–14.10.

According to Figure 14.8 plain slot-entry bearings yield stiffness values at least as good as, if not better than, recessed bearings. The stiffness of recessed hydrostatic bearings is reduced by

Table 14.1: Hydrostatic stiffness, hydrodynamic stiffness, and squeeze damping for thin-land recessed journal bearings

$\lambda_{hs} = \dfrac{P_s LD}{h_o}\dfrac{3n^2}{2\pi}\sin^2(\pi/n)\dfrac{(1-a/L)\beta}{Z+1+2\gamma\sin^2(\pi/n)}$	Hydrostatic stiffness
$\lambda_{hd} = \dfrac{P_s LD}{h_o}12n^2\sin^2(\pi/n)\,S_h\cdot\dfrac{(a/L)(L/D)^2(1-a/L)^2}{Z+1+2\gamma\sin^2(\pi/n)}$	Hydrodynamic stiffness
$C_{sq} = \dfrac{\lambda_{hd}}{\pi N}$	Squeeze damping
$\gamma = \dfrac{n}{\pi}\dfrac{a}{b}\left(1-\dfrac{a}{L}\right)\dfrac{L}{D}$	Circumferential flow factor
$Z = \beta/(1-\beta)$	Capillary, hole or slot control
$Z = 0.5\beta/(1-\beta)$	Orifice or turbulent control
$Z = 0$	Constant flow control

Table 14.2: Hydrostatic stiffness, hydrodynamic stiffness, and squeeze damping for plain line-entry journal bearings

$$\lambda_{hs} = \frac{P_s LD}{h_o} \beta \frac{3n^2}{2\pi} \sin^2 (\pi/n) \frac{F_1}{F_2}$$

Hydrostatic stiffness

$$\lambda_{hd} = \frac{P_s LD}{h_o} 2n^2 \sin^2 (\pi/n) S_h \left(\frac{L}{D}\right)^2 \cdot F_3$$

Hydrodynamic stiffness

$$C_{sq} = \frac{P_s LD}{h_o} 2\frac{n^2}{\pi} \sin^2 (\pi/n) S_o \left(\frac{L}{D}\right)^2 \cdot F_3 = \frac{\lambda_{hd}}{\pi N}$$

Squeeze damping

$$F_1 = (1 - a/L) - \frac{1}{3}(L/D)^2 \cdot F_4$$

$$F_2 = Z + 1 + 2\frac{a}{L}\left(1 - \frac{a}{L}\right)\left(\frac{L}{D}\right)^2$$

$$F_3 = \frac{a}{L}\left(1 - \frac{2a}{L}\right)^2 \frac{6}{F_2} + F_4$$

$$F_4 = \frac{(a/L)^3}{1 + \frac{1}{3}\left(\frac{a}{L}\right)^2 \left(\frac{L}{D}\right)^2} + \frac{(1 - 2a/L)^3}{1 + \frac{1}{3}\left(1 - \frac{2a}{L}\right)^2 \left(\frac{L}{D}\right)^2}$$

$$S_h = \frac{\eta N}{P_s}\left(\frac{D}{2h_o}\right)^2 \qquad S_o = \frac{\eta}{P_s}\left(\frac{D}{2h_o}\right)^2$$

Z as in Table 14.1

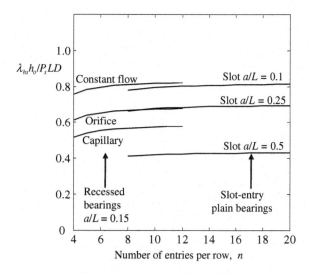

Figure 14.8: Variation of Concentric Hydrostatic Stiffness with Number of Feed-Entry Positions Around a Journal Bearing, $L/D = 1$, $\beta = 0.5$.

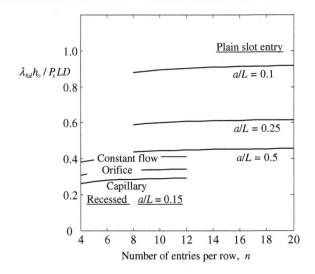

Figure 14.9: Variation of Concentric Hydrodynamic Stiffness with Number of Feed-Entry Positions Around a Journal Bearing, $L/D = 1$, $K = 1$, $\beta = 0.5$.

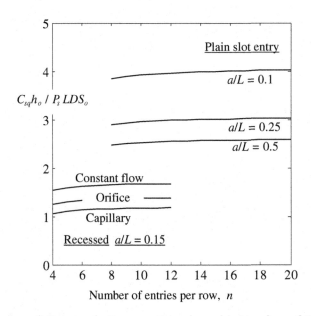

Figure 14.10: Variation of Concentric Squeeze Damping with Number of Feed-Entry Positions Around a Journal Bearing, $L/D = 1$, $\beta = 0.5$.

circumferential flow. The values shown in the figure are for a circumferential flow land 20% larger than the axial flow land.

Figure 14.9 shows the substantial advantage of typical plain bearing configurations for increased hydrodynamic stiffness. Figure 14.10 shows the substantial advantage of typical plain bearing configurations for increased squeeze film damping.

Rigid Mass Supported Concentrically on Hydrostatic Bearings

The shaft in Figure 14.6 is rigid, and associated with each bearing there is a mass m. The equations of motion, ignoring compressibility, may be written as

$$\begin{aligned} m\ddot{x} + C_{sq}\dot{x} + \lambda_{hs}x - \lambda_{hd}y &= 0 \\ m\ddot{y} + C_{sq}\dot{y} + \lambda_{hs}y + \lambda_{hd}x &= w \end{aligned} \tag{14.41}$$

These equations may be normalized and solved by the methods described in Section 14.6. The equations are normalized in terms of the characteristic frequency $\omega_n = \sqrt{\lambda_{hs}/m}$ and the damping ratio $\zeta = C/C_c$, where

$$\varsigma = \frac{C_{sq}\omega_n}{2\lambda_{hs}} = \frac{C_{sq}/2}{\sqrt{\lambda_{hs}m}}$$

The amplitude and phase results are

$$\frac{X}{Y} = \frac{\lambda_{hd}/\lambda_{hs}}{\sqrt{A^2 + B^2}}; \qquad \phi_1 = -\tan^{-1}(B/A)$$

$$\lambda_{hs}\frac{Y}{W} = \sqrt{\frac{A^2 + B^2}{C^2 + D^2}}; \qquad \phi_2 = \phi_1 - \tan^{-1}(D/C) \tag{14.42}$$

$$\lambda_{hs}\frac{X}{W} = \frac{\lambda_{hd}/\lambda_{hs}}{\sqrt{C^2 + D^2}}; \qquad \phi_3 = -\tan^{-1}(D/C)$$

where

$$A = 1 - \left(\frac{\omega}{\omega_n}\right)^2$$

$$B = 2\varsigma\frac{\omega}{\omega_n}$$

$$C = \left[1 - \left(\frac{\omega}{\omega_n}\right)^2\right]^2 - \left(2\varsigma\frac{\omega}{\omega_n}\right)^2 + \left(\frac{\lambda_{hd}}{\lambda_{hs}}\right)^2$$

$$D = 4\varsigma\frac{\omega}{\omega_n}\left[1 - \left(\frac{\omega}{\omega_n}\right)^2\right]$$

Example 14.3 Rigid Mass Supported Concentrically in Hydrostatic Journal Bearings

Determine the hydrostatic stiffness λ_{hs}, hydrodynamic stiffness λ_{hd}, and damping C_{sq} for the following 12-slot line-entry data. Also determine the natural frequency ω_n, damping ratio ζ ($= C/C_c$) and vibration amplitudes at various speeds of rotation for a small-amplitude sinusoidal force applied in the y direction.

Given data:

Supply pressure $P_s = 4$ MN/m^2 (580 lbf/in^2)
Bearing length $L = 50$ mm (2.0 in)
Bearing diameter $D = 50$ mm (2.0 in)
Bearing film thickness $h_o = 50$ μm (0.002 in)
Speed when $K = 1$ $N_o = 13.33$ rev/s, the speed at which
                         $K = 1$, $S_{ho} = 0.073$

Axial land width $\alpha = 5$ mm (0.02 in)
Concentric pressure ratio $\beta = 0.5$
Rigid mass $m = 101$ kg.

Solution
From Figure 14.8:

$$\lambda_{hs} = \frac{4 \times 10^6 \times 50 \times 50 \times 10^{-6}}{50 \times 10^{-6}} \times 0.8 = 160 \text{ MN/m } (0.914 \times 10^6 \text{ lbf/in})$$

$$\lambda_{hd} = \frac{4 \times 10^6 \times 50 \times 50 \times 10^{-6}}{50 \times 10^{-6}} \times 0.9 = 180 \text{ MN/m } (1.03 \times 10^6 \text{ lbf/in})$$

$$C_{sq} = \frac{180 \times 10^6}{\pi \times 13.33} = 4.3 \text{ MN s/m } (24,600 \text{ lbf s/in})$$

$$\omega_n = \sqrt{\frac{\lambda_{hs}}{m}} = \sqrt{\frac{160 \times 10^6}{101}} = 1260 \text{ rad/s}$$

This corresponds to a frequency $f_n = 200$ Hz and a ratio $f_n/N_o = 200/13.33 = 15$. The high ratio means the natural frequency is above and well removed from the operating speed.

The damping ratio $\zeta = C_{sq}\omega_n/2\lambda_{hs}$ and since $C_{sq} = \lambda_{hd}/\pi N$:

$$\varsigma = \frac{\lambda_{hd}}{\lambda_{hs}}\frac{\omega_n}{2\pi N} = \frac{0.9}{0.8}\frac{1260}{2 \times \pi \times 13.33} = 16.9$$

This is a very high damping value and means the response at zero rotational speed will decay to a very low value.

Noting that λ_{hd} increases with N, a range of operating conditions may be summarized as in Table 14.3. The frequency responses may be evaluated using the values in Table 14.3 and the expressions in equation (14.42). The calculated values of the responses in the direction of excitation are shown in Figure 14.11. There are also responses in the x direction and at high rotational speeds these will often be considerably larger than the vibrations in the y direction. This is particularly true for analysis of responses when the journal is in the concentric condition and gives rise to the possibility of whirl instability.

It can be seen from Figure 14.11 that the responses at low speed are very heavily damped. However, as the rotational speed is increased the resonant amplitudes start to increase up to a critical speed when the amplitude approaches infinity. This is an indication of whirl instability, a condition discussed further below.

■

Table 14.3: Operating conditions for Example 14.3

N (rev/s)	N/N_o	$\overline{\lambda}_{hs}$	$\overline{\lambda}_{hd}$	$\overline{\lambda}_{hd}/\overline{\lambda}_{hs}$	ζ
0	0	0.8	0	0	16.9
13.33	1	0.8	0.9	1.125	16.9
53.32	4	0.8	3.6	4.5	16.9
213.3	16	0.8	14.4	18	16.9
400	30	0.8	27	33.75	16.9

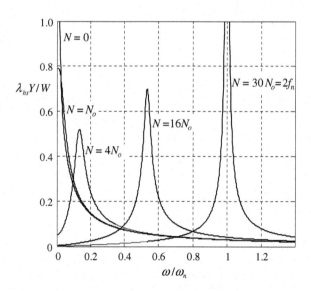

Figure 14.11: Amplitude of Vibration in the Direction of Excitation Against Excitation Frequency for Example 14.3.

Whirl Instability

Whirl instability is illustrated in Figure 14.11 for a case where the rotational speed of the shaft is $N = 30N_o$. The speed in this case is for a natural frequency $f_n = 200$ Hz. The natural frequency gives rise to a rotational speed ratio $f_n/N = 0.5$. This type of whirl instability is therefore known as "half-speed whirl" because when the journal rotates at twice the natural frequency, the journal starts to orbit around the bearing center of rotation at half the rotational speed. In other words, the journal orbits at the natural frequency because at this whirl speed the hydrodynamic effect cancels out the squeeze damping effect. The consequence is similar to the response of a simple spring—mass system with zero damping. The condition that gives rise to "half-speed whirl" is explained in the following few paragraphs.

The puzzling question is: *How is it possible for a heavily damped system to act as though it has zero damping?* The explanation is not altogether simple and lies in the nature of the hydrodynamic stiffness. At high rotational speeds, hydrodynamic stiffness is much higher than hydrostatic stiffness. Application of a small force in the y direction causes a response movement in the x direction due to the hydrodynamic stiffness rather than a movement in the y direction under the influence of the hydrostatic stiffness. As the velocity occurs in the x direction, a squeeze damping force reacts against the shaft in the x direction, tending to dissipate the energy of vibration. This is fine for a low-speed stable bearing. Next, consider the situation of a whirling shaft.

Whirl is when the center of the shaft orbits around the bearing following an approximately circular locus. When the shaft is displaced in the y direction, the velocity of the shaft center is in the x direction. Similarly, when the shaft is displaced along the x direction, the velocity is maximal along the y direction. The effect of the hydrodynamic stiffness is to cause a positive reaction force in the direction of the velocity. This force acts in opposition to the squeeze force and cancels out the damping.

The hydrodynamic stiffness therefore gives rise in the way just described to negative damping for whirl motion. The net damping is reduced to zero when the whirling speed of the shaft center is half the rotational speed of the shaft. When, in addition, the whirling speed coincides with the natural frequency, the response tends to become infinite and the bearing is unstable. This occurs when the rotational speed N approaches $2f_n$, which is termed the onset speed for "half-speed whirl". It is impossible to increase the speed to run through whirl. The result would be the destruction of the machine.

Techniques are available to increase the rotational speed at which instability occurs. One method is to operate with high eccentricity ratio due to the steady load. This works by reducing the attitude angle between the steady load and the response. Another method is to introduce an additional damping medium between the bearing housing and the bearing. Rubber O-rings have been employed for this purpose.

Half-speed whirl should not be confused with "synchronous whirl" due to shaft unbalance. Unbalance causes whirl of the shaft center at rotational speed. Obviously a resonance occurs as the rotational speed approaches a natural frequency. Synchronous whirl is not instability, although damage may occur if the unbalance is too great. There may be several resonant frequencies, which are usually termed "critical speeds" of the system.

14.10 Eccentric Journal Bearings

As mentioned above, the effect of increasing eccentricity ratio at first reduces stability and then, at high eccentricity ratio, improves stability. These effects can be analyzed from knowledge of bearing stiffness and damping coefficients, as explained in greater detail for hybrid bearings by Rowe et al. (1984) and by Rowe and Chong (1986).

Definitions given in Sections 14.8 and 14.9 are helpful in understanding the physical meaning of the coefficients. Two direct and two cross-coupling stiffness coefficients are expressed in terms of dimensionless factors as

$$a_{ij} = \frac{P_s LD}{h_o} \, \bar{a}_{ij} \tag{14.43}$$

The meanings of the terms a_{ij} can be seen from equations (14.28) and (14.29). The damping terms are expressed as

$$b_{ij} = \frac{P_s LD}{\omega_o h_o} \, \bar{b}_{ij} \tag{14.44}$$

The symbol $\omega_o = 2\pi N_o$ represents the angular rotational speed at which $K = 1$ and the corresponding value of speed parameter is S_{ho}.

Typical variations of stiffness and damping coefficients are shown in Figure 14.12 for a 12-slot plain bearing, $a/L = 0.1$, $\beta = 0.5$, $L/D = 1$, $S_{ho} = 0.073$, and $K = 3$. It is seen that stiffness and damping coefficients all increase with eccentricity ratio. However, the increase in hydrodynamic stiffness is destabilizing at higher speeds, as explained in previous paragraphs. This is further demonstrated by a drop in the speed threshold for instability, as seen in a typical example for $K = 3$ (Figure 14.13). At high values of eccentricity ratio, the speed threshold increases again due to a reduction in the attitude angle defined in Figure 14.7.

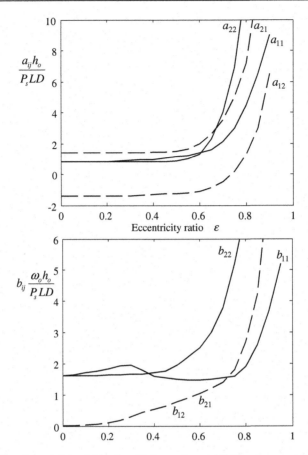

Figure 14.12: Stiffness and Damping Coefficients for a 12-Slot Journal Bearing: $a/L = 0.1$, $L/D = 1.0$, $\beta = 0.5$, $K = 3$, $S_{ho} = 0.073$.

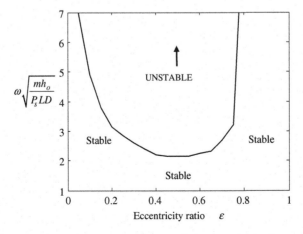

Figure 14.13: Whirl Speed Threshold for a Rigid 12-Slot Journal Bearing: $a/L = 0.1$, $L/D = 1.0$, $\beta = 0.5$, $K = 3$, $S_{ho} = 0.073$, $\omega = 2\pi N$.

The speed threshold parameter is further reduced as shaft flexibility is increased (see Rowe et al., 1984).

References

Rowe, W. B. (1980). Dynamic and static properties of recessed hydrostatic journal bearings by small displacement analysis. *Journal of Lubrication Technology, Transactions of the American Society of Mechanical Engineers, 102*(1; Jan.), 71–79.

Rowe, W. B., & Chong, F. S. (1986). Computation of dynamic force coefficients for hybrid (hydrostatic/hydrodynamic) journal bearings by the finite disturbance and perturbation techniques. *Tribology International, 19*(5; Oct.), 260–271.

Rowe, W. B., Weston, W., & Koshal, D. (1979). Static and dynamic properties of concentric hydrostatic journal bearings by small displacement analysis. *Euromech Colloquium.* No. 124, Fiat Research Centre, Orbassano, Italy, 2–4 October 1979.

Rowe, W. B., Chong, F. S., & Weston, W. (1984). A linearized stability analysis of rigid and flexible rotors in plain hybrid (hydrostatic/hydrodynamic) journal bearings. *Proceedings of the Institution of Mechanical Engineers,* 277–286, Paper C262/84.

CHAPTER 15

Experimental Methods and Testing

15.1 Introduction

The practical assessment of a bearing if carried out correctly will reveal whether the actual bearing performance matches the specification. This provides a check on previous calculation, design, and manufacture. In practice, if problems are experienced, these are more likely to arise due to bearing parts that lie outside manufacturing tolerances than due to any other causes. It is obviously unrealistic, for example, to expect measured flow to lie within the specified range if the bearing clearance is half or double the specified value. The starting point for practical assessment of a bearing is therefore to make a thorough dimensional investigation of the machine and its parts before and after final assembly. Some features to look for are:

- Sizes that govern bearing clearance
- Parallelism of shafts, sliders, and their mating surfaces
- Squareness of shafts and thrust flanges
- Flatness of sliding parts
- Roundness of rotating parts
- Position and sizes of recesses
- Position and size of bearing land widths
- Position and size of flow restrictors.

If measured values are different from specified values, it may still be possible to predict performance of the bearing as made and hence decide whether it is necessary to correct the machine faults or whether it is possible to proceed.

A further useful test before final assembly is to measure flow through the restrictors when the lubricant is supplied at supply pressure. This checks on the size of the restrictors, but also checks on free flow through each restrictor and further on the pressure ratio. If flow through a restrictor outside the machine is q_1 and flow through the assembled machine is q_2, the pressure ratio is given by

$$\beta = \frac{q_1 - q_2}{q_1}$$

An alternative method to find q_1 is to measure flow with the shaft or slider removed from the machine so that there is no bearing film resistance to impede the flow.

Hydrostatic, Aerostatic and Hybrid Bearing Design.
DOI: 10.1016/B978-0-12-396994-1.00015-2

Subsequent to dimensional metrology and restrictor calibration, the machine should be carefully assembled in a clean environment to prevent the ingress of solid particles that may subsequently lead to wear or restrictor blockage. The system should then be carefully flushed by pumping an appropriate lubricant through the bearing, through the filter, and through the rest of the system for a sufficient period to eliminate any loose particles that may have escaped earlier detection. Flow rate may then be checked again to ensure no reduction due to blockage of a restrictor. All flow measurements need to be carried out under the same stable temperature conditions.

By this stage the free-running nature of the machine will have been ascertained. In the absence of any rubbing seals, the machine should move absolutely smoothly without the slightest hesitation.

Further investigation may proceed to determine such performance qualities as:

- Variations of bearing film thickness with load
- Flow variations with bearing film thickness
- Friction
- Temperature rise
- Bearing pressures.

Measurement of these qualities has been discussed in detail in previous publications (see Section 15.4). The following sections illustrate the use of two particular types of rig that have given reliable load results for flat pads and cylindrical journal bearings.

15.2 Flat-Pad Rig

Single thrust pads of various shapes may be tested in rigs of the type shown in Figures 15.1 and 15.2. The flat pad in both rigs is loaded by air pressure applied to a piston guided in frictionless air bearings. The attractive feature of this type of test arrangement is the ease of ensuring alignment of the test surfaces.

The pad to be tested is bolted onto the lower face of the test rig. A displacement transducer mounted on the upper surface of the fixing plate senses movement of the loading piston and by this means is used to measure thickness of the bearing film.

Various displacement transducers have been successfully employed for bearing test work, including transducers based on eddy currents, inductance, and capacitance. The efficiency of such transducers depends, amongst other things, on the electrical properties of the materials used in the construction of the opposing surfaces.

The steady load applied to the pad is measured by means of a pressure transducer, which indicates the pressure on the top face of the loading piston. The total load is the pressure force plus the weight of the loading piston.

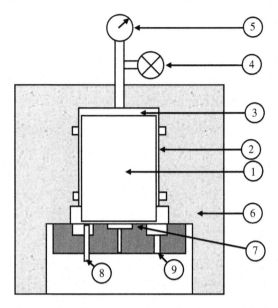

Figure 15.1: Static Loading Representation of a Structure Including Bearings.

Figure 15.2: A Flat-Pad Test Rig.

Referring to Figure 15.1, the center shaft (1) with its axis vertical is located radially in two gas-lubricated journal bearings (2). The lower end of the shaft forms one surface of the bearing film. The chamber (3) at the top of the shaft is fed by air through the pressure regulator (4) and the pressure in the chamber is monitored by a pressure transducer or gauge (5).

Adjacent to the lower end of the shaft (1), a ground and lapped thrust plate (7) forms the other surface of the test bearing.

The force exerted by the gas pressure in the chamber (3) on the end of the shaft plus the weight of the shaft forms the total load on the bearing. The pneumatic chamber acts as a decoupling spring between the shaft and the rig main structure (6).

Three position sensors (8) indicate the gap between the shaft and the test bearing. Particular care is taken in the construction of the rig to ensure squareness of the lower face of the main structure and the surface of the test bearing and the surface of the test shaft.

Drainage from the test bearing is provided by outlets (9) that feed back to the tank of the lubricant pressure supply system. An actual test rig is shown in Figure 15.2. Further refinements were introduced to provide for both static and dynamic loading.

15.3 Cylindrical Journal Bearing Rig

The main features of a cylindrical journal bearing test rig are illustrated in Figures 15.3 and 15.4. Essentially this rig contains a precision shaft (1), within which are mounted two miniature transducers. The two transducers consist of a displacement transducer and a pressure transducer, both mounted with extreme care to avoid disturbance to the lubricant film.

The shaft is rotated by a drive motor (2) and traversed axially through the test bearing (3) by means of a slideway and a lead screw (4). The motions allow pressure and film thickness to be measured at any point on the bearing circumference under rotating or stationary conditions. Additional transducers (5) mounted on the end faces of the test bearing assist in the measurement of shaft eccentricity and alignment. A steady load W is applied by means of tightening a vertical screw attached to a lead screw through a decoupling spring (6) and a force transducer (7).

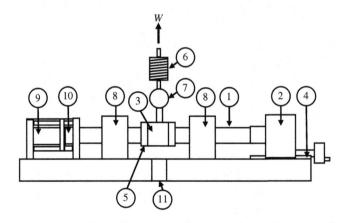

Figure 15.3: Schematic Layout of a Journal Bearing Test Rig.

Figure 15.4: A Journal Bearing Test Rig.

Applied load on the bearing is reacted through the shaft (1) and the support bearings (8). The support bearings themselves are hydrostatic to offer the smoothest running and maximum load support. The support bearings, besides being longer than the test bearing, are supplied with much higher pressure.

Signals from transducers inside the shaft are taken out by means of slip-rings (9). The shaft is connected to the drive motor and the slip-rings at each end by means of flexible couplings (10).

The test bearing would be free to move in three axes of translation and two axes of rotation if no additional constraints were provided. Figure 15.4 shows the test bearing constrained by an alignment jig. The alignment jig is designed on the basis of location through tensioned wires. The principle of the tensioned wire is that maximum force is required to obtain movement along the axis of the wire and minimum force for very small movement normal to the axis. Thus, the wires are positioned so as to allow the bearing to move radially with respect to the shaft but not in any other direction. Other attachments to the bearing include the decoupled loading mechanism and the flexible oil supply pipe. Care is essential to avoid appreciable constraint due to the oil supply pipe. For friction measurements, the null position method is employed to determine the torque required to maintain a zero angular deflection. Leakage from the bearing is allowed to fall to the base plate and flow back to the supply tank through the drain pipe (11).

Flow measurements simply require a volume of oil to be collected from the drain over a measured period of time under stable operating conditions, thus giving the results shown in Figure 15.5 for a four-recess hydrostatic bearing. It is seen that flow increases by approximately 15% with increasing eccentricity ratio. Experimental results for an inter-recess land-width angle of $\theta = 30°$ are given as points compared with a continuous theoretical line. The experimental results are more than 4% greater than the predicted values. Significant errors can occur due to variations in lubricant temperature since viscosity tends to be strongly dependent on temperature. Small errors in bearing film thickness also have a strong effect due to the cubic relationship, and thermal expansion causes a difference between measured clearance and the clearance under experimental conditions for the flow measurements. Another possible source of experimental error is the control of supply pressure. While it would be sensible for a critical parameter to strive for greater accuracy of agreement between theory and experiment, the general level of agreement in this case was considered to be acceptable.

Load results for a 12-slot double-row plain journal bearing are presented in Figure 15.6 for a wide range of speeds ranging from zero speed to a speed corresponding to a power ratio $K = 119$. The load results are in close agreement with predictions for zero speed $K = 0$ and for $K = 1.36$ across the range of values of eccentricity ratio. This is despite the difficulty of maintaining accurate alignment between the shaft and the bearing at high values of eccentricity ratio. For higher values of power ratio, there are larger experimental errors at high eccentricity ratios, as would be expected. Experimental results can exceed predicted values at very high values of eccentricity ratio exceeding approximately 0.8 at the zero-speed condition, where there is no hydrodynamic load support. This was identified as being due to a collapse of

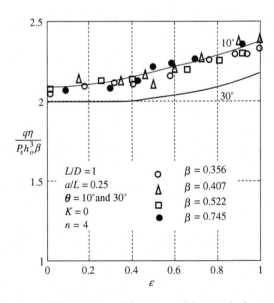

Figure 15.5: Variation of Flow Rate with Eccentricity Ratio in Hydrostatic Journal Bearing Experiments.

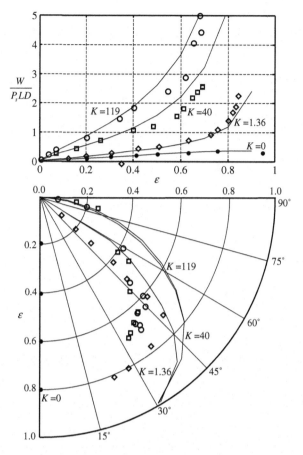

Figure 15.6: Load and Attitude Angle with Eccentricity Ratio for Hydrostatic Double-Entry Slot Journal Bearings: $L/D = 1$, $a/L = 0.1$, $\beta = 0.19$, $S_{ho} = 0.045$.

bearing film stiffness with the consequence that the shaft takes partial load support from metal-to-metal contact. As the bearing load is reduced, full fluid-film support is resumed and experimental results revert to the predicted trend.

Attitude angle results are also presented in Figure 15.6. These show a significant overestimation of attitude angle compared with measured values except at high values of power ratio and low values of eccentricity ratio. A significant source of error at high experimental values of eccentricity ratio occurs due to the difficulty of maintaining zero misalignment of the journal within the bearing. Any misalignment causes near metal-to-metal contact at high eccentricity ratio. This has the effect of making the bearing film appear to carry higher loads than is possible with a perfectly aligned bearing and tends to reduce attitude angle. Another source of error in prediction of attitude angle is the difficulty of accurately representing cavitation boundaries. The most substantial errors occur at high values of eccentricity ratio, tending to support cavitation as a source of errors. In the prediction technique used for the results shown, no account was taken of whether there was sufficient fluid available within the bearing to support the hydrodynamic pressures predicted. Other results using more sophisticated techniques achieved limited success in gaining improved agreement with experiments.

Experiments are essential for validation of a design but the conclusion reached is always that it is far easier to predict what should happen than to accurately measure what actually does happen. Despite this conclusion, it is surprisingly easy to construct a simple rig to determine whether a basic design works.

15.4 Publications on Experimental Behavior

The literature on hydrostatic and aerostatic bearing behavior is very extensive. The Southampton University Gas Bearing Symposia were for some years a notable source of information. The proceedings of professional bodies with their specialized journals on lubrication technology and tribology are also notable.

The References section below lists some relevant publications by the author concerning bearing applications described in this book. The papers therein all include experimental techniques and experimental data. Rowe (1967) describes experiments on four grinding machine spindles, including aerostatic, hydrostatic, and hydrodynamic spindles. Rowe and O'Donoghue (1970) consider the development and testing of diaphragm valve-controlled hydrostatic bearings. Leonard and Davies (1971) study the dynamic behavior of a four-recess hydrostatic journal bearing. O'Donoghue et al. (1971) detail experiments on a Yates bearing. Rowe and Kilmister (1974) describe experiments on a compliant surface aerostatic bearing. Stout et al. (1974) cover experiments on slot-restricted journal bearings. Rowe et al. (1976, 1977) describe experiments on recessed and plain hybrid journal bearings. Rowe et al. (1980) consider the development and testing of a bearing for the Daresbury Ion Beam Accelerator. Koshal and Rowe (1980, 1981) describe further experiments on hybrid bearings. Chang et al. (1986) document experiments comparing dynamic results from sinusoidal testing and from impact testing of journal bearings. Other experimental work supervised by the author is contained in Ph.D. theses published by his valued co-authors, who were mostly research students and postdoctoral fellows.

References

Chang, K. K., Chong, F. S., & Rowe, W. B. (1986). Journal bearing vibration — a comparison of sinusoidal excitation with impact excitation. *Proceedings of the 1st European Turbo-Machinery Symposium*. Institution of Mechanical Engineers. October.

Koshal, D., & Rowe, W. B. (1980). Fluid-film bearings operating in a hybrid mode, Parts 1 and 2. Transactions of the ASME. *Journal of Lubrication Technology, 103*, 558–572.

Koshal, D., & Rowe, W. B. (1981). Fluid-film journal bearings operating in a hybrid mode: Part II — Experimental investigation. *Journal of Lubrication Engineering, Transactions of the American Society of Mechanical Engineers, 103*(3; Oct.), 566–573.

Leonard, R., & Davies, P. B. (1971). An experimental investigation of the dynamic behaviour of a four-recess hydrostatic journal bearing. *Proceedings of the Conference on Externally Pressurized Bearings.* Institution of Mechanical Engineers. London, pp. 245–261.

O'Donoghue, J. P., Wearing, R. S., & Rowe, W. B. (1971). Multi-recess externally pressurized bearings using the Yates principle. *Proceedings of the Conference on Externally Pressurized Bearings.* Institution of Mechanical Engineers. London, pp. 337–351.

Rowe, W. B. (1967). Experience with four types of grinding machine spindles. *Proceedings of the 8th International Machine Tool Design and Research Conference.* September, pp. 453–457. Oxford: Pergamon Press.

Rowe, W. B., & Kilmister, G. T. F. (1974). A theoretical and experimental investigation of a self-compensating externally pressurized thrust bearing. *Proceedings of the 6th International Gas Bearing Symposium.* Southampton University. March, D1, pp. 1–13.

Rowe, W. B., & O'Donoghue, J. P. (1970). Diaphragm valves for controlling opposed pad hydrostatic bearings. Tribology Convention, Brighton. *Proceedings of the Institution of Mechanical Engineers.*

Rowe, W. B., Koshal, D., & Stout, K. J. (1976). Slot-entry bearings for hybrid hydrodynamic and hydrostatic operation. *Journal of Mechanical Engineering Science, 18*(2), 73–78.

Rowe, W. B., Koshal, D., & Stout, K. J. (1977). Investigation of recessed and slot-entry journal bearings for hybrid hydrodynamic and hydrostatic operation. *Wear, 43*(1; May), 55–70.

Rowe, W. B., Morris, M. C., & Acton, J. (1980). Hydrostatic bearing for the Daresbury Laboratory. *Tribology International,* 183–184, August.

Stout, K. J., Porritt, T. E., & Rowe, W. B. (1974). The performance of externally pressurized slot-restricted bearings. *Proceedings of the 6th International Gas Bearing Symposium.* Southampton University. March, Paper A6, pp. 83–96.

Index

Note: Page numbers followed by "*f*" and "*t*" denote figures and tables, respectively.